CAMBRIDGE LIBRARY COLLECTION

Books of enduring scholarly value

Darwin

Two hundred years after his birth and 150 years after the publication of 'On the Origin of Species', Charles Darwin and his theories are still the focus of worldwide attention. This series offers not only works by Darwin, but also the writings of his mentors in Cambridge and elsewhere, and a survey of the impassioned scientific, philosophical and theological debates sparked by his 'dangerous idea'.

Erasmus Darwin

The author of this life of Erasmus Darwin (1731–1802), published in 1879, is given as Ernst Krause (1839–1903), a German biologist, but in fact more than half the book is a 'preliminary notice' by Erasmus's grandson Charles Darwin, who explains in the preface that he has written it because of his access to family papers which add 'to the knowledge of Erasmus Darwin's character'. Krause wrote his original article in a German periodical because, in turn, he was intrigued by a reference made by Charles Darwin in the later editions of On the Origin of Species to his grandfather's anticipation in his Zoonomia (also reissued in this series) of Lamarckian theory: 'I thought immediately ... that this ancestor of his must certainly deserve considerable credit in connection with the history of the Darwinian theory.' The German text was translated by W.S. Dallas, who had previously collaborated with Darwin as both indexer and translator.

Cambridge University Press has long been a pioneer in the reissuing of out-of-print titles from its own backlist, producing digital reprints of books that are still sought after by scholars and students but could not be reprinted economically using traditional technology. The Cambridge Library Collection extends this activity to a wider range of books which are still of importance to researchers and professionals, either for the source material they contain, or as landmarks in the history of their academic discipline.

Drawing from the world-renowned collections in the Cambridge University Library, and guided by the advice of experts in each subject area, Cambridge University Press is using state-of-the-art scanning machines in its own Printing House to capture the content of each book selected for inclusion. The files are processed to give a consistently clear, crisp image, and the books finished to the high quality standard for which the Press is recognised around the world. The latest print-on-demand technology ensures that the books will remain available indefinitely, and that orders for single or multiple copies can quickly be supplied.

The Cambridge Library Collection will bring back to life books of enduring scholarly value (including out-of-copyright works originally issued by other publishers) across a wide range of disciplines in the humanities and social sciences and in science and technology.

Erasmus Darwin

ERNST KRAUSE
CHARLES ROBERT DARWIN

CAMBRIDGE
UNIVERSITY PRESS

CAMBRIDGE UNIVERSITY PRESS

Cambridge, New York, Melbourne, Madrid, Cape Town, Singapore,
São Paolo, Delhi, Dubai, Tokyo

Published in the United States of America by Cambridge University Press, New York

www.cambridge.org
Information on this title: www.cambridge.org/9781108006163

© in this compilation Cambridge University Press 2009

This edition first published 1879
This digitally printed version 2009

ISBN 978-1-108-00616-3 Paperback

ERASMUS DARWIN.

FROM A PICTURE BY WRIGHT OF DERBY

ERASMUS DARWIN.

By ERNST KRAUSE.

TRANSLATED FROM THE GERMAN BY W. S. DALLAS.

WITH A PRELIMINARY NOTICE

By CHARLES DARWIN.

PORTRAIT AND WOODCUTS.

LONDON:

JOHN MURRAY, ALBEMARLE STREET.

1879.

PREFACE.

—◦—

In the February number, 1879, of a well-known German scientific journal, 'Kosmos,' Dr. Ernst Krause published a sketch of the life of Erasmus Darwin, the author of the ' Zoo-nomia,' ' Botanic Garden,' and other works. This article bears the title of a ' Contribution to the history of the Descent-Theory ; ' and Dr. Krause has kindly allowed my brother Erasmus and myself to have a translation made of it for publication in this country.*

As I have private materials for adding to the knowledge of Erasmus Darwin's character, I have written a preliminary notice. These materials consist of a large collection of letters written by him ; of his common-place book in folio, in the possession of his grandson Reginald Darwin ; of some notes made shortly

---

* Mr. Dallas has undertaken the translation, and his scientific reputation, together with his knowledge of German, is a guarantee for its accuracy.

after his death, by my father, Dr. Robert
Darwin, together with what little I can
clearly remember that my father said about
him; also some statements by his daughter,
Violetta Darwin, afterwards Mrs. Tertius
Galton, written down at the time by her
daughters; and various short published notices.
To these must be added the 'Memoirs of the
Life of Dr. Darwin,' by Miss Seward, which
appeared in 1804; and a lecture by Dr. Dowson
on "Erasmus Darwin, Philosopher, Poet,
and Physician," published in 1861, which
contains many useful references and remarks.*

---

* Since the publication of Dr. Krause's article, Mr. Butler's
work, 'Evolution, Old and New, 1879,' has appeared, and this
includes an account of Dr. Darwin's life, compiled from the two
books just mentioned, and of his views on evolution.

PRELIMINARY NOTICE.

Erasmus Darwin was descended from a Lincolnshire family, and the first of his ancestors of whom we know anything was William Darwin, who possessed a small estate at Cleatham.* He was also yeoman of the armoury of Greenwich to James I. and Charles I. This office was probably almost a sinecure, and certainly of very small value. He died in 1644, and we have reason to believe from gout. It is, therefore, probable that Erasmus, as well as many other members of the family, inherited from this William, or some of his predecessors, their strong tendency to gout; and it was an early attack of gout which made Erasmus a vehement advocate for temperance throughout his whole life.

* The greater part of the estate of Cleatham was sold in 1760. A cottage with thick walls, some fish-ponds and old trees, alone show where the "Old Hall" once stood. A field is still called the "Darwin Charity," from being subject to a charge, made by the second Mrs. Darwin, for buying gowns for four old widows every year.

B

The second William Darwin (born 1620) served as Captain-Lieutenant in Sir W. Pelham's troop of horse, and fought for the king. His estate was sequestrated by the Parliament, but he was afterwards pardoned on payment of a heavy fine. In a petition to Charles II. he speaks of his almost utter ruin from having adhered to the royal cause, and it appears that he had become a barrister. This circumstance probably led to his marrying the daughter of Erasmus Earle, Serjeant-at-law; and hence Erasmus Darwin derived his Christian name.

The eldest son from this marriage, William (born 1655), married the heiress of Robert Waring, of Wilsford, in the county of Nottingham. This lady also inherited the manor of Elston, which has remained ever since in the family.

This third William Darwin had two sons —William, and Robert who was educated as a barrister, and who was the father of Erasmus. I suppose that the Cleatham and the Waring properties were left to William, who seems to have followed no profession, and the Elston estate to Robert; for when

Elston Hall (where Erasmus Darwin was born), as it existed before 1754.

the latter married, he gave up his profession
and lived ever afterwards at Elston.   There is
a portrait of him at Elston Hall, and he looks,
with his great wig and bands, like a dignified
doctor of divinity. He seems to have had some
taste for science, for he was an early member
of the well-known Spalding Club; and the
celebrated antiquary, Dr. Stukeley, in ' An
account of the almost entire Sceleton of a large
animal,' &c., published in the 'Philosophical
Transactions,' April and May 1719, begins
his paper as follows :—" Having an account
" from my friend, Robert Darwin, Esq., of
" Lincoln's Inn, a Person of Curiosity, of a
" human Sceleton impressed in Stone, found
" lately by the Rector of Elston," &c. Stukeley
then speaks of it as a great rarity, " the like
" whereof has not been observed before in this
" island, to my knowledge."   Judging from
a sort of litany written by Robert, and handed
down in the family, he was a strong advocate
of temperance, which his son ever afterwards
so strongly advocated :—

> From a morning that doth shine,
> From a boy that drinketh wine,
> From a wife that talketh Latine,
> Good Lord deliver me.

It is suspected that the third line may be accounted for by his wife, the mother of Erasmus, having been a very learned lady.

The eldest son of Robert, christened Robert Waring, succeeded to the estate of Elston, and died there at the age of ninety-two, a bachelor. He had a strong taste for poetry, like his youngest brother Erasmus. Robert also cultivated botany, and when an oldish man, he published his 'Principia Botanica.' This book in MS. was beautifully written, and my father declared that he believed it was published because his old uncle could not endure that such fine calligraphy should be wasted. But this was hardly just, as the work contains many curious notes on biology—a subject wholly neglected in England in the last century. The public, moreover, appreciated the book, as the copy in my possession is the third edition.

Of the second son, William Alvey, I know nothing. A third son, John, became the rector of Elston, the living being in the gift of the family. The fourth son, and the youngest of the children, was Erasmus, the subject of the present memoir, who was born on the 12th Dec. 1731, at Elston Hall.

His elder brother, Robert, states, in a letter to my father (May 19, 1802), that Erasmus " was always fond of poetry. He was also " always fond of mechanicks. I remember " him when very young making an ingenious " alarum for his watch (clock ?); he used also " to show little experiments in electricity " with a rude apparatus he then invented " with a bottle." The same tastes, therefore, appeared very early in life which prevailed to the day of his death. " He had always a " dislike to much exercise and rural diver- " sions, and it was with great difficulty that " we could ever persuade him to accompany " us."

When ten years old (1741), he was sent to Chesterfield School, where he remained for nine years. His sister, Susannah, wrote to him at school in 1748, and I give part of the letter as a curiosity. She was then a young lady between eighteen and nineteen years old. She died unmarried, and her nephew, Dr. Robert Darwin (my father), who was deeply attached to her, always spoke of her as the very pattern of an old lady, so nice looking, so gentle, kind, and charitable, and passionately fond of flowers. The first part of her letter consists of gossip and family news, and is not worth giving.

Susannah Darwin *to* Erasmus.

Dear Brother,

I come now to yᵉ chief design of my Letter, and that is to acquaint you with my Abstinence this Lent, which you will find on yᵉ other side, it being a strict account of yᵉ first 5 days, and all yᵉ rest has been conformable thereto ; I shall be glad to hear from you wᵗʰ an account of your temperance this lent, wᶜʰ I expect far exceeds mine. As soon as we kill our hog I intend to take part thereof with yᵉ Family, for I'm informed by a learned Divine yᵗ Hogs Flesh is Fish, and has been so ever since yᵉ Devil entered into yᵐ and they ran into yᵉ Sea ; if you and the rest of the Casuists in your neighbourhood are of yᵉ same oppinion, it will be a greater satisfaction to me, in resolving so knotty a point of Conscience. This being all at present I conclude with all our dues to you and Broʳ.

Your affectionate sister,
S. Darwin.

A Diary in Lent.

Elston, Feb. 20, 1748.

Febʳʸ 8 Wednesday Morning a little before seven I got up ; said my Prayers ; worked till eight ;

y^n took a walk, came in again and eate a farthing Loaf, y^n dress'd me, red a Chapter in ye Bible, and spun till One, y^n dined temperately viz: on Puddin, Bread and Cheese ; spun again till Fore, took a walk, y^n spun till half an hour past Five ; eat an Apple, Chattered round ye Fire ; and at Seven a little boyl'd Milk; and y^n (takeing my leave of Cards ye night before) spun till nine; drank a Glass of Wine for ye Stomack sake ; and at Ten retired into my Chamber to Prayers ; drew up my Clock and set my Larum betwixt Six and Seven.

Thursday call'd up to Prayers, by my Larum ; spun till Eight, collected ye Hens' Eggs ; breakfasted on Oat Cake, and Balm Tea; y^n dress'd and spun till One, Pease Porrage, Pottatoes and Apple Pye ; y^n turned over a few pages in Scribelerus ; eat an Apple and got to my work ; at Seven got Apple Pye and Milk, half an hour after eight red in ye Tatlar and at Ten withdrew to Prayers; slept sound; rose before Seven; eat a Pear ; breakfast a quarter past Eight ; fed ye Cats, went to Church ; at One Pease Porrage, Puddin, Bread and Cheese ; Fore Mrs. Chappells came, Five drank Tea ; Six eat half an Apple ; Seven a Porrenge of Boyl'd Milk ; red in ye Tatlar ; at Eight a Glass of Punch ; filled up ye vacancies of ye day with work as before.

Saturday Clock being too slow lay rather longar y^n usal; said my Prayers; and breakfasted at Eight ; at One broth, Pudding, Brocoli and Eggs, and

Apple Pye; at Five an Apple; seven Apple Pye, Bread and Butter; at Nine a Glass of Wine; at Ten Prayers.

Sunday breakfast at Eight; at Ten went to ye Chappell; 12 Dumplin, red Herring, Bread and Cheese; two to ye Church; read a Lent Sermon at Six; and at Seven Appel Pye Bread and Cheese.

Excuse hast, being very cold.

ERASMUS, ÆTAT. 16, *to* SUSANNAH DARWIN.

DEAR SISTER,

I receiv'd yours about a fortnight after ye date y^t I must begg to be excused for not answering it sooner: besides I have some substantial Reasons, as having a mind to see Lent almost expired, before I would vouch for my Abstinence throughout ye whole: and not having had a convenient oppertunity to consult a Synod of my learned friends about your ingenious Conscience, and I must inform you we unanimously agree in ye Opinion of ye Learned Divine you mention, that Swine may indeed be fish but then they are a devillish sort of fish; and we can prove from ye same Authority that all fish is flesh whence we affirm Porck not only to be flesh but a devillish Sort of flesh; and I would advise you for Conscience sake altogether to abstain from tasting it; as I can assure You I have done, tho' roast Pork has come to Table several Times; and for my own part

have lived upon Puding, milk, and vegetables all this
Lent; but don't mistake me, I don't mean I have
not touch'd roast beef, mutton, veal, goose, fowl, &c.
for what are all these? All flesh is grass! Was I
to give you a journal of a Week, it would be stuft so
full of Greek and Latin as translation Verses, themes,
annotation Exercise and ye like, it would not only be
very tedious and insipid but perfectly unintelligible
to any but Scholboys.

I fancy you forgot in Yours to inform me y^t
your Cheek was quite settled by your Temperance,
but however I can easily suppose it. For ye temperate enjoy an ever-blooming Health free from all y^e
Infections and disorders luxurious mortals are subject
to, the whimsical Tribe of Phisitians cheated of their
fees may sit down in penury and Want, they may
curse mankind and imprecate the Gods and call down
y^t parent of all Deseases, luxury, to infest Mankind,
luxury more distructive than ye Sharpest Famine;
tho' all the Distempers that ever Satan inflicted upon
Job hover over ye intemperate; they would play
harmless round our Heads, nor dare to touch a single
Hair. We should not meet those pale thin and
haggard countenances which every day present themselves to us. No doubt men would still live their
Hunderd, and Methusalem would lose his Character;
fever banished from our Streets, limping Gout would
fly ye land, and Sedentary Stone would vanish into
oblivion and death himself be slain.

I could for ever rail against Luxury, and for ever panegyrize upon abstinence, had I not already encroach'd too far upon your Patience, but it being Lent the exercise of y^t Christian virtue may not be amiss, so I shall proceed a little furder—

[The remainder of the letter is hardly legible or intelligible, with no signature.]

P.S.—Excuse Hast, supper being called, very Hungry.

Judging from two letters—the first written in 1749, to one of the under-masters during the holidays, and the other to the headmaster, shortly after he went to Cambridge, in 1750—he seems to have felt a degree of respect, gratitude, and affection for the several masters unusual in a schoolboy. Both these letters were accompanied by an inevitable copy of verses, those addressed to the head-master being of considerable length, and in imitation of the 5th Satire of Persius. His two elder brothers accompanied him to St. John's College, Cambridge ; and this seems to have been a severe strain on their father's income. They appear, in consequence, to have been thrifty and honourably economi-

cal; so much so that they mended their own
clothes; and, many years afterwards, Erasmus
boasted to his second wife that, if she cut the
heel out of a stocking, he would put a new one
in without missing a stitch. He won the
Exeter Scholarship at St. John's, which was
worth only £16 per annum. No doubt he
studied the classics whilst at Cambridge,
for he did so to the end of his life, as shown
by the many quotations in his latest work,
'The Temple of Nature.' He must also have
studied mathematics to a certain extent, for,
when he took his Bachelor of Arts degree, in
1754, he was at the head of the Junior Optimes.
Nor did he neglect medicine; and he left
Cambridge during one term to attend Hunter's
lectures in London. As a matter of course,
he wrote poetry whilst at Cambridge, and a
poem on ' The Death of Prince Frederick,' in
1751, was published many years afterwards, in
1795, in the European Magazine.

In the autumn of 1754 he went to Edinburgh to study medicine, and while there,
seems to have been as rigidly economical as at
Cambridge; for amongst his papers there is a
receipt for his board from July 13th to October

13th, amounting to only £6 12s. Mr. Keir, afterwards a distinguished chemist, was at Edinburgh with him, and after his death wrote to my father (May 12th, 1802) : " The classical " and literary attainments which he had ac- " quired at Cambridge gave him, when he " came to Edinburgh, together with his poeti- " cal talents and ready wit, a distinguished " superiority among the students there. " Every one of the above-mentioned Pro- " fessors [whose lectures he attended], except- " ing Dr. Whytt, had been a pupil of the " celebrated Boerhaave, whose doctrines were " implicitly adopted. It would be curious to " know (but he alone could have told us) the " progress of your father's mind from the " narrow Boerhaavian system, in which man " was considered as an hydraulic machine " whose pipes were filled with fluid suscep- " tible of chemical fermentations, while the " pipes themselves were liable to stoppages " or obstructions (to which obstructions and " fermentations all diseases were imputed), " to the more enlarged consideration of man " as a *living being*, which affects the phenomena " of health and disease more than his merely

" mechanical and chemical properties. It is
" true that about the same time, Dr. Cullen
" and other physicians began to throw off the
" Boerhaavian yoke; but from the minute
" observation which Dr. Darwin has given
" of the laws of association, habits and phe-
" nomena of animal life, it is manifest that his
" system is the result of the operation of his
" own mind."

The only other record of his life in Edinburgh which I possess is a letter to his friend
Dr. Okes, of Exeter,* written shortly after
the death of his father (1754), when he was
twenty-three years old. It shows his sceptical
frame of mind whilst he was quite a young
man.

ERASMUS DARWIN *to* DR. OKES.

"Yesterday's post brought me the disagreeable
news of my father's departure out of this sinful
world.

" He was a man of more sense than learning; of
very great industry in the law, even after he had no
business, nor expectation of any. He was frugal,
but not covetous; very tender to his children, but

* Published by one of his descendants in the ' Gentleman's
Magazine,' Oct. 1808, vol. lxxviii. pt. ii. p. 869.

still kept them at an awful kind of distance. He passed through this life with honesty and industry, and brought up seven healthy children to follow his example.

"He was 72 years old, and died the 20th of this current November 1754. 'Blessed are they that die in the Lord.'

"That there exists a superior ENS ENTIUM, which formed these wonderful creatures, is a mathematical demonstration. That HE influences things by a particular providence, is not so evident. The probability, according to my notion, is against it, since general laws seem sufficient for that end. Shall we say no particular providence is necessary to roll this Planet round the Sun, and yet affirm it necessary in turning up *cinque* and *quatorze*, while shaking a box of dies? or giving each his daily bread? The light of Nature affords us not a single argument for a future state; this is the only one, that it is possible with God, since He who made us out of nothing can surely re-create us; and that He will do this is what we humbly hope. I like the Duke of Buckingham's epitaph—'Pro Rege sæpe, pro Republicâ semper, dubius, non improbus vixi; incertus, sed inturbatus morior. Christum advenero, Deo confido benevolenti et omnipotenti, Ens Entium miserere mei!'

"ERASMUS DARWIN."

The expression "disagreeable news," applied to his father's death, sounds very odd to our ears, but he evidently used this word where we should say "painful." For, in a feeling letter to Josiah Wedgwood, the famous potter, written a quarter of a century afterwards (Nov. 29th, 1780), about the death of their common friend Bentley, in which he alludes to the death of his own son, he says nothing but exertion will dispossess "the " *disagreeable* ideas of our loss."

In 1755 he returned to Cambridge, and took his Bachelor of Medicine degree. He then again went to Edinburgh, and early in Sept. 1756, settled as a physician in Nottingham. Here, however, he remained for only two or three months, as he got no patients. Whilst in Nottingham he wrote several letters, some in Latin and some in English, to his friend, the son of the famous German philosopher, Reimarus.* Mechanics and medicine were the bonds of union between them. Erasmus also dedicated a poem to young Reimarus, on his taking his degree at Leyden

---

* I am much indebted to a son of Dr. Sieveking, who brought to England the original letters preserved by the descendants of Reimarus, for permitting me to have them photographed.

in 1754. Various subjects were discussed between them, including the wildest speculations by Erasmus on the resemblance between the action of the human soul and that of electricity, but the letters are not worth publishing. In one of them he says: "I believe "I forgot to tell how Dr. Hill makes his "'Herbal' (a formerly well-known book). "He has got some wooden plates from some "old herbal, and the man that cleans them "cuts out one branch of every one of them, "or adds one branch or leaf, to disguise "them. This I have from my friend Mr. "G——y, watch-maker, to whom this print- "mender told it, adding, 'I make plants now "every day that God never dreamt of.'" It also appears from one of his letters to Reimarus, that Erasmus corresponded at this time about short-hand writing with Gurney, the author of a well-known book on this subject. Whilst still young he filled six volumes with short-hand notes, and continued to make use of the art for some time.

Several of the letters to Reimarus relate to a case in which Dr. Darwin appears to have been much interested. He sent or helped to

send a working man to a London surgeon,
Mr. D., for a serious operation. Reimarus
and Dr. Darwin appear to have had some
misunderstanding with the surgeon, expect-
ing that he would perform the operation
gratuitously. Dr. Darwin writes to Reimarus :
" I am very sorry to hear that D. took six
" guineas from the poor young man. He
" has nothing but what hard labour gives
" him ; is much distressed by this thing
" costing him near £30 in all, since the
" house where he lay cheated him much.
"   .   .   .   When he returns I shall send
" him two guineas. I beg you would not
" mention to my brother that I send this
" to him." Why his brother should not be
told of this act of charity it is difficult to
conjecture. From two other letters it appears
that Dr. Darwin wrote anonymously to his
friend the surgeon, complaining of his charge ;
and that when suspected of this discreditable
act he did not own the authorship of the
letter. He wrote to Reimarus (Nottingham
Sept. 9th, 1756) : " You say I am suspected
" to be the Author of it (*i.e.* the anonymous
" letter), and next to me some malicious per-

" son somewhere else, and that I am desired
" as I am a gentleman to declare concerning
" it. First, then, as I am upon Honour, I
" must not conceal that I am glad there are
" Persons who will revenge Faults the Law
" can not take hold off: and I hope Mr. D.
" will not be affronted at this Declaration;
" since you say he did not know the Distress
" of the Man. Secondly, as another Person
" is suspected, I will not say whether I am
" the Author or not, since I don't think the
" Author merits Punishment, for informing
" Mr. D. of a Mistake. You call the Letter
" a threatening Letter, and afterwards say
" the Author pretends to be a Friend to
" Mr. D. This, though you give me several
" particulars of it, is a Contradiction I don't
" understand." In a P.S. he adds that Reimarus might show the letter to Mr. D. The
anonymous letter answered its purpose, for
the surgeon returned four guineas, and Dr.
Darwin thought it probable that he would
ultimately return the other two guineas.

In November 1756, Erasmus settled in
Lichfield, and now his life may be said to

have begun in earnest; for it was here, and
in or near Derby, to which place he removed
in 1781, that he published all his works.
Owing to two or three very successful cases,
he soon got into some practice at Lichfield as
a physician, when twenty-five years old. A
year afterwards (Dec. 1757) he married Miss
Mary Howard, aged 17–18 years, who, judging
from all that I have heard of her, and from
some of her letters, must have been a superior
and charming woman. She died after a long
and suffering illness in 1770. They seem to
have lived together most happily during the
thirteen years of their married life, and she
was tenderly nursed by her husband during her
last illness. Miss Seward gives,* on second-
hand authority, a long speech of hers, ending
with the words, " he has prolonged my days,
" and he has blessed them." This is probably
true, but everything which Miss Seward says
must be received with caution; and it is
scarcely possible that a speech of such
length could have been reported with any
accuracy.

The following letter was written by

* 'Memoirs of the Life of Dr. Darwin,' 1804, pp. 11–14.

Erasmus four days before his marriage with
Miss Howard.

ERASMUS DARWIN *to* MARY HOWARD.

DEAR POLLY,   DARLASTON, *Dec.* 24, 1757.

As I was turning over some old mouldy
volumes, that were laid upon a Shelf in a Closet of
my Bed-chamber; one I found, after blowing the
Dust from it with a Pair of Bellows, to be a Receipt
Book, formerly, no doubt, belonging to some good
old Lady of the Family. The Title Page (so much
of it as the Rats had left) told us it was "a Bouk off
verry monny muckle vallyed Receipts bouth in
Kookery and Physicks." Upon one Page was "To
make Pye-Crust,"—in another "To make Wall-
Crust,"—"To make Tarts,"—and at length "To
make Love." "This Receipt," says I, "must be
curious, I'll send it to Miss Howard next Post, let
the way of making it be what it will."—Thus it is
"To make Love. Take of Sweet-William and of
Rose-Mary, of each as much as is sufficient. To the
former of these add of Honesty and Herb-of-grace;
and to the latter of Eye-bright and Motherwort of
each a large handful: mix them separately, and
then, chopping them altogether, add one Plumb,
two sprigs of Heart's Ease and a little Tyme. And
it makes a most excellent dish, probatum est. Some
put in Rue, and Cuckold-Pint, and Heart-Chokes,

and Coxcome, and Violents; But these spoil the flavour of it entirely, and I even disprove of Sallery which some good Cooks order to be mix'd with it. I have frequently seen it toss'd up with all these at the Tables of the Great, where no Body would eat of it, the very appearance was so disagreable."

Then follow'd "Another Receipt to make Love," which began "Take two Sheep's Hearts, pierce them many times through with a Scewer to make them Tender, lay them upon a quick Fire, and then taking one Handful——" here Time with his long Teeth had gnattered away the remainder of this Leaf. At the Top of the next Page, begins "To make an honest Man." "This is no new dish to me," says I, "besides it is now quite old Fashioned; I won't read it." Then follow'd "To make a good Wife." "Pshaw," continued I, "an acquaintance of mine, a young Lady of Lichfield, knows how to make this Dish better than any other Person in the World, and she has promised to treat me with it sometime," and thus in a Pett threw doun the Book, and would not read any more at that Time. If I should open it again tomorrow, whatever curious and useful receipts I shall meet with, my dear Polly may expect an account of them in another Letter.

I have the Pleasure of your last Letter, am glad to hear thy cold is gone, but do not see why it should keep you from the concert, because it was gone. We drink your Health every day here, by

the Name of Dulcinea del Toboso, and I told Mrs.
Jervis and Miss Jervis that we were to have been
married yesterday, about which they teased me all
the Evening. I heard nothing of Miss Fletcher's
Fever before. I will certainly be with Thee on
Wednesday evening, the Writings are at my House,
and may be dispatched that night, and if a License
takes up any Time (for I know nothing at all about
these Things) I should be glad if Mr. Howard would
order one, and by this means, dear Polly, we may
have the Ceremony over next morning at eight
o'clock, before any Body in Lichfield can know
almost of my being come Home. If a License is to
be had the Day before, I could wish it may be put
off till late in the Evening, as the Voice of Fame
makes such quick Dispatch with any News in so
small a Place as Lichfield.—I think this is much
the best scheme, for to stay a few Days after my
Return could serve no Purpose, it would only make
us more watch'd and teazed by the Eye and Tongue
of Impertinence.—I shall by this Post apprize my
Sister to be ready, and have the House clean, and I
wish you would give her Instructions about any
trivial affairs, that I cannot recollect, such as a cake
you mentioned, and tell her the Person of whom,
and the Time when it must be made, &c. I'll desire
her to wait upon you for this Purpose. Perhaps
Miss Nelly White need not know the precise Time
till the Night before, but this as you please, as
I (*illegible*). You could rely upon her Secrecy, and

it's a Trifle, if any Body should know. Matrimony, my dear Girl, is undoubtedly a serious affair, (if any Thing be such) because it is an affair for Life: But, as we have deliberately determin'd, do not let us be *frighted* about this Change of Life; or however, not let any breathing Creature perceive that we have either Fears or Pleasures upon this Occasion: as I am certainly convinced, that the best of Confidants (tho' experienced on a thousand other Occasions) could as easily hold a burning cinder in their Mouth as anything the least ridiculous about a new married couple! I have ordered the Writings to be sent to Mr. Howard that he may peruse and fill up the blanks at his Leizure, as it wilt (I foresee) be dark night before I get to Lichfield on Wednesday. Mrs. Jervis and Miss desire their Compl. to you, and often say how glad she shall be to see you for a few Days at any Time. I shall be glad, Polly, if thou hast Time on Sunday night, if thou wilt favour me with a few Lines by the return of the Post, to tell me how Thou doest, &c.—My Compl. wait on Mr. Howard if He be returned.—My Sister will wait upon you, and I hope, Polly, Thou wilt make no Scruple of giving her Orders about whatever you chuse, or think necessary. I told her Nelly White is to be Bride-Maid. Happiness attend Thee! adieu

from, my dear Girl,

thy sincere Friend,

E. DARWIN.

P.S.—Nothing about death in this Letter, Polly.

It has been said that he soon got into practice at Lichfield, and I have found the following memorandum of his profits in his own
handwriting :—

The profits of my business amounted

				£	s.	d.
From Nov. 12, 1756 to Jan. 1, 1757				18	7	6
Jan.	1757	„	1758	192	10	6
„	1758	„	1759	305	2	0
„	1759	„	1760	469	4	0
„	1760	„	1761	544	2	0
„	1761	„	1762	669	18	0
„	1762	„	1763	726	0	0
From Jan. 12, 1763 to Jan. 1, 1764				639	13	0
„	1764	„	1765	750	13	0
„	1765	„	1766	800	1	4
„	1766	„	1767	748	5	6
„	1767	„	1768	847	3	0
„	1768	„	1769	775	11	6
„	1769	„	1770	?		
„	1770	„	1771	956	17	6
„	1771	„	1772	1064	7	6
„	1772	„	1773	1025	3	0

Later in life he gave up the good habit of
keeping accurate accounts, for in 1799 he wrote
to my father that he had been much perplexed
what return to make to the commissioners
(of income tax ?), as " I kept no book, but be-
" lieved my business to be £1000 a year, and de-

" duct £200 for travelling expenses and chaise
" hire, and £200 for a livery-servant, four
" horses and a day labourer." Subsequently
he informed my father that the commissioners
had accepted this estimate. A century ago
an income of £1000 would probably be equal
to one of £2000 at the present time; but I am
greatly surprised that his profits were not
larger. All his friends constantly refer to
his long and frequent journeys, for his practice lay chiefly amongst the upper classes
of society. When he went to live at the
Priory, he remarked to my father in a letter
that five or six additional miles would make
little difference in the fatigue of his journeys.

In 1781, eleven years after the death of
his first wife, he married the widow of
Colonel Chandos Pole, of Radburn Hall. He
had become acquainted with her in the
Spring of 1778, when she had come to
Lichfield in order that he might attend her
children professionally. It is evident from
the many MS. verses addressed to her before
their marriage, that Dr. Darwin was passionately attached to her, even during the life-
time of her husband, who died in 1780.

These verses are somewhat less artificial than his published ones. On his second marriage he left Lichfield, and after living two years at Radburn Hall, he removed into the town of Derby, and ultimately to Breadsall Priory, a few miles from the town, where he died in 1802.

There is little to relate about his life at either Lichfield or Derby, and, as I am not attempting a connected narrative, I will here give such impressions as I have formed of his intellect and character, and a few of his letters which are either interesting in themselves, or which throw light upon what he thought and felt.

His correspondence with many distinguished men was large; but most of the letters which I possess or have seen are uninteresting, and not worth publication. Medicine and mechanics alone roused him to write with any interest. He occasionally corresponded with Rousseau, with whom he became acquainted in an odd manner, but none of their letters have been preserved. Rousseau was living in 1766 at Mr. Davenport's house, Wootton Hall, and used to

spend much of his time " in the well-known
" cave upon the terrace in melancholy con-
" templation." He disliked being interrupted,
so Dr. Darwin, who was then a stranger to
him, sauntered by the cave, and minutely
examined a plant growing in front of it. This
drew forth Rousseau, who was interested in
botany, and they conversed together, and
afterwards corresponded during several years.

I find a letter written in February 1767 on
a singular subject. A gentleman had consulted him about the body of an infant which
had apparently been murdered. It was
believed to be the illegitimate child of a
lady, and to have been murdered by its
mother. He kept a copy of this letter,
without any address. Omitting all medical
details it runs as follows :—

DEAR SIR,   LICHFIELD, *Feb.* 7, 1767.

I am sorry you should think it necessary to
make any excuse for a Letter I this morning received
from you. The Cause of Humanity needs no Apology
to me.

    *   *   *   *   *   *

The Women that have committed this most
unnatural crime, are real objects of our greatest

Pity; their education has produced in them so much Modesty, or sense of Shame, that this artificial Passion overturns the very instincts of Nature!—what Struggles must there be in their minds, what agonies!—at a Time when, after the Pains of Parturition, Nature has designed them the sweet Consolation of giving Suck to a little helpless Babe, that depends on them for its hourly existence!—Hence the cause of this most horrid crime is an excess of what is really a Virtue, of the Sense of Shame, or Modesty. Such is the Condition of human Nature! I have carefully avoided the use of scientific terms in this Letter that you may make any use of it you may think proper; and shall only add that I am veryly convinced of the Truth of every part of it.

and am, Dear Sir,

Your affectionate friend and servant,

ERASMUS DARWIN.

There is, perhaps, no safer test of a man's real character than that of his long continued friendship with good and able men. Now, Mr. Edgeworth, the father of Maria Edgeworth, the authoress, asserts,* after mentioning the names of Keir, Day, Small, Bolton, Watt, Wedgwood, and Darwin, that "their "mutual intimacy has never been broken

---

* 'Memoirs of R. L. Edgeworth,' 2nd ed. vol. i. p. 181.

" except by death." To these names, those
of Edgeworth himself and of the Galtons
may be added. The correspondence in my
possession shows the truth of the above
assertion.   Mr. Day was a most eccentric
character, whose life has been sketched by
Miss Seward: he named Erasmus Darwin
" as one of the three friends from whom he
" had met with constant kindness;"* and
Dr. Darwin, in a letter to my father, says:
" I much lament the death of Mr. Day.
" The loss of one's friends is one great evil
" of growing old.   He was dear to me by
" many names (*multis mihi nominibus charus*),
" as friend, philosopher, scholar, and honest
" man."

I give below two of his letters to Josiah
Wedgwood.

ERASMUS DARWIN *to* JOSIAH WEDGWOOD.

DEAR WEDGWOOD,   LICHFIELD, *Sept.* 30, 1772.

I did not return soon enough out of Derby-
shire to answer your letter by yesterday's Post.
Your second letter gave me great consolation about
Mrs. Wedgewood, but gave me most sincere grief

* ' Memoirs of R. L. Edgeworth,' 2nd ed. vol. ii. p. 113.

about Mr. Brindley, whom I have always esteemed
to be a great genius, and whose loss is truly a public
one. I don't believe he has left his equal. I think
the various Navigations should erect him a monument in Westminster Abbey, and hope you will at
the proper time give them this hint.

Mr. Stanier sent me no account of him, except of
his death, though I so much desired it, since if I had
understood that he got worse, nothing should have
hindered me from seeing him again. If Mr. Hen-
shaw took any Journal of his illness or other circumstances after I saw him, *I wish you would ask him for
it and enclose it to me.* And any circumstances that
you recollect of his life should be wrote down, and I
will some time digest them into an Eulogium. These
men should not die, this Nature denies, but their
Memories are above her Malice. Enough!

*　　*　　*　　*　　*

ERASMUS DARWIN *to* JOSIAH WEDGWOOD.

DEAR SIR,   LICHFIELD, *Nov.* 29, 1780.

Your letter communicating to me the death of
your friend, and I beg I may call him mine Mr.
Bentley, gives me very great concern; and a train of
very melancholy ideas succeeds in my mind, uncon-
nected indeed with your loss, but which still at times
casts a shadow over me, which nothing but exertion
in business or in acquiring knowledge can remove.

This exertion I must recommend to you, as it for a
time dispossesses the disagreeable ideas of our loss;
and gradually their impression or effect upon us
becomes thus weakened, till the traces are scarcely
perceptible, and a scar only is left, which reminds us
of the past pain of the united wound.

Mr. Bentley was possessed of such variety of knowledge, that his loss is a public calamity, as well as
to his friends, though they must feel it the most
sensibly! Pray pass a day or two with me at Lichfield, if you can spare the time, at your return. I
want much to see you; and was truly sorry I was
from home as you went up; but I do beg you will
always lodge at my house on your road, as I do at
yours, whether you meet with me at home or not.

I have searched in vain in Melmoth's translation
of Cicero's letters for the famous consolatory letter
of Sulpicius to Cicero on the loss of his daughter (as
the work has no index), but have found it, the first
letter in a small publication called 'Letters on the
most common as well as important occasions in Life:'
Newberry, St. Paul's, 1758. This letter is a masterly
piece of oratory indeed, adapted to the man, the time,
and the occasion. I think it contains everything
which could be said upon the subject, and if you have
not seen it I beg you to send for the book.

For my own part, too sensible of the misfortunes of
others for my own happiness, and too pertinacious of
the remembrance of my own [i.e. the death of his

son Charles in 1778], I am rather in a situation to demand than to administer consolation. Adieu. God bless you, and believe me, dear Sir, your affectionate friend

E. DARWIN.

Ten years later he seems to have doubted much about the consolation to be derived from the letter of Sulpicius, for he writes (1790) to Edgeworth :*

I much condole with you on your late loss. I know how to feel for your misfortune. The little Tale you sent is a prodigy, written by so young a person, with such elegance of imagination. *Nil admirari* may be a means to escape misery, but not to procure happiness. There is not much to be had in this world—we expect too much! I have had my loss also. The letter of Sulpicius to Cicero is fine eloquence, but comes not to the heart; it tugs, but does not draw the arrow. Pains and diseases of the mind are only cured by Time. Reason but skins the wound, which is perpetually liable to fester again.

Amongst the old letters preserved, there is one without any date from Hutton, the founder of the modern science of geology, and I extract its commencement, as proceeding from so illustrious a scientific man. Dr.

* 'Memoirs,' 2nd ed. 1821, vol. ii. p. 110.

Darwin seems to have complained to him of having been cheated by some publisher; and Hutton answers :—

If you have no more money than you use, then be as sparing of it as you please, but if you have money to spend, then pray learn to let yourself be cheated, that is, learn to lay out money for which you have no other use. If this be not philosophy, at least it is good sense; for why the devil should a man have money to be a plague to him, when it is so easy to throw it away; and if thro' a spirit of general benevolence you are afraid of mankind suffering from this root of all evil, for God's sake send it to the bottom of the sea, it there can only poison fish and it will there make in time a noble fossil specimen.

One of his granddaughters has remarked to me, that the term "benevolent" has been associated with his name, almost in the same manner as that of "judicious" with the name of the old divine, Hooker. This is perfectly true, for I have incessantly met with this expression in letters and in the many published notices about him. To the word benevolent, sympathy is generally added, and often generosity, as well as hospitality. Mr. Edgeworth says : * " I have known him intimately

* 'Monthly Magazine' 1802 p. 115.

" during thirty-six years, and in that period
" have witnessed innumerable instances of his
" benevolence."

His life-long friend, Mr. Keir, wrote to my
father (May 12th, 1802) about his character
as follows: " I think all those who knew
" him, will allow that sympathy and be-
" nevolence were the most striking features.
" He felt very sensibly for others, and, from
" his knowledge of human nature, he entered
" into their feelings and sufferings in the
" different circumstances of their constitution,
" character, health, sickness, and prejudice.
" In benevolence, he thought that almost all
" virtue consisted. He despised the monkish
" abstinences and the hypocritical pretensions
" which so often impose on the world. The
" communication of happiness and the relief
" of misery were by him held as the only
" standard of moral merit. Though he ex-
" tended his humanity to every sentient
" being, it was not like that of some philo-
" sophers, so diffused as to be of no effect;
" but his affection was there warmest where
" it could be of most service to his family
" and his friends, who will long remember

" the constancy of his attachment and his
" zeal for their welfare." His neighbour, Sir
Brooke Boothby, after the loss of his child (to
whom the beautiful and well-known monument in Ashbourne church was erected), in
an ode addressed to Dr. Darwin, writes in
strong terms about his sympathy and power
of consolation.

But it is fair to state that from my father's
conversation, I infer that Dr. Darwin had
acted towards him in his youth rather
harshly and imperiously, and not always
justly; and though in after years he felt the
greatest interest in his son's success, and
frequently wrote to him with affection,
in my opinion the early impression on
my father's mind was never quite obli-
terated.

I have heard indirectly (through one of his
stepsons) that he was not always kind to his
son Erasmus, being often vexed at his retiring
nature, and at his not more fully display-
ing his great talents. On the other hand
his children by his second marriage seem to
have entertained the warmest affection for
him.

ERASMUS DARWIN *to* HIS SON ROBERT.

DEAR ROBERT, *April* 19, 1789.

I am sorry to hear you say you have many enemies, and one enemy often does much harm. The best way, when any little slander is told one, is never to make any piquant or angry answer; as the person who tells you what another says against you, always tells them in return what you say of them. I used to make it a rule always to receive all such information very coolly, and never to say anything biting against them which could go back again; and by these means many who were once adverse to me, in time became friendly. Dr. Small always went and drank tea with those who he heard had spoken against him; and it is best to show a little attention at public assemblies to those who dislike one; and it generally conciliates them.

*　　*　　*　　*　　*

Robert seems to have consulted his father about some young man, whom he wished to see well started as an apothecary, and received the following answer :—

ERASMUS DARWIN *to* HIS SON ROBERT.

DEAR ROBERT, DERBY, *Dec.* 17, 1790.

I cannot give any letters of recommendation to Lichfield, as I am and have been from their in-

fancy acquainted with all the apothecaries there;
and as such letters must be directed to some of their
patients, they would both feel and resent it. When
Mr. Mellor went to settle there from Derby I took
no part about him. As to the prospect of success
there, if the young man who is now at Edinburgh
should take a degree (which I suppose is probable),
he had better not settle in Lichfield.

I should advise your friend to use at first all
means to get acquainted with the people of all ranks.
At first a parcel of blue and red glasses at the windows
might gain part of the retail business on market
days, and thus get acquaintance with that class of
people. I remember Mr. Green, of Lichfield, who is
now growing very old, once told me his retail business, by means of his show-shop and many-coloured
window, produced him £100 a year. Secondly, I
remember a very foolish, garrulous apothecary at
Cannock, who had great business without any knowledge or even art, except that he persuaded people
he kept good drugs; and this he accomplished by
only one stratagem, and that was by *boring* every
person who was so unfortunate as to step into his
shop with the goodness of his drugs. " Here's a fine
piece of assafœtida, smell of this valerian, taste this
album græcum. Dr. Fungus says he never saw such
a fine piece in his life." Thirdly, dining every
market day at a farmers' ordinary would bring him
some acquaintance, and I don't think a little impedi-

ment in his speech would at all injure him, but
rather the contrary by attracting notice. Fourthly,
card assemblies,—I think at Lichfield surgeons are
not admitted as they are here;—but they are to
dancing assemblies; these therefore he should attend.
Thus have I emptied my quiver of the *arts* of the
Pharmacopol. Dr. K——d, I think, supported his
business by perpetual boasting, like a Charlatan;
this does for a blackguard character, but ill suits a
more polished or modest man.

If the young man has any friends at Shrewsbury
who could give him letters of introduction to the
proctors, this would forward his getting acquaintance. For all the above purposes some money must
at first be necessary, as he should appear well;
which money cannot be better laid out, as it will
pay the greatest of all interest by settling him
well for life. Journeymen Apothecaries have not
greater wages than many servants; and in this state
they not only lose time, but are in a manner lowered
in the estimation of the world, and less likely to
succeed afterwards. I will certainly send to him,
when first I go to Lichfield. I do not think his
impediment of speech will injure him; I did not find
it so in respect to myself. If he is not in such
narrow circumstances but that he can appear well,
and has the knowledge and sense you believe him
to have, I dare say he will succeed anywhere. A
letter of introduction from you to Miss Seward, men-

tioning his education, may be of service to him, and
another from Mr. Howard. Adieu, from, dear
Robert,

> Yours most affectionately,
>
>   E. DARWIN.

My father spoke of Dr. Darwin as having
great powers of conversation. Lady Charle-
ville, who had been accustomed to the most
brilliant society in London, told him that Dr.
Darwin was one of the most agreeable men
whom she had ever met. He himself used
to say "there were two sorts of agreeable
" persons in conversation parties—agreeable
" talkers and agreeable listeners."

He stammered greatly, and it is surprising
that this defect did not spoil his powers
of conversation. A young man once asked
him in, as he thought, an offensive manner,
whether he did not find stammering very
inconvenient. He answered, "No, Sir, it
" gives me time for reflection, and saves me
" from asking impertinent questions." Miss
Seward speaks of him as being extremely
sarcastic, but of this I can find no evidence
in his letters or elsewhere. It is a pity that
Dr. Johnson in his visits to Lichfield rarely

met Dr. Darwin; but they seem to have
disliked each other cordially, and to have
felt that if they met they would have quar-
relled like two dogs. There can, I suppose,
be little doubt that Johnson would have come
off victorious. In a volume of MSS. by
Dr. Darwin, in the possession of one of his
granddaughters, there is the following stanza:

> From Lichfield famed two giant critics come,
> Tremble, ye Poets! hear them! "Fe, Fo, Fum!"
> By Seward's arm the mangled Beaumont bled,
> And Johnson grinds poor Shakespear's bones for bread.

He is evidently alluding to Mr. Seward's
edition of 'Beaumont and Fletcher's Plays,'
and to Johnson's edition of 'Shakespear'
in 1765.

He possessed, according to my father, great
facility in explaining any difficult subject; and
he himself attributed this power to his habit
of always talking about whatever he was
studying, "turning and moulding the subject
" according to the capacity of his hearers."
He compared himself to Gil Blas's uncle, who
learned the grammar by teaching it to his
nephew.

When he wished to make himself disagreeable for any good cause, he was well able to do so. Lady * * * married a widower, and became so jealous of his former wife that she cut and spoiled her picture, which hung up in one of the rooms. The husband, fearing that his young wife was becoming insane, was greatly alarmed, and sent for Dr. Darwin. When he arrived he told her in the plainest manner many unpleasant truths, amongst others that the former wife was infinitely her superior in every respect, including beauty. The poor lady was astonished at being thus treated, and could never afterwards endure his name. He told the husband if she again behaved oddly, to hint that he would be sent for. The plan succeeded perfectly, and she ever afterwards restrained herself.

My father was much separated from Dr. Darwin after early life, so that he remembered few of his remarks, but he used to quote one saying as very true: " that the " world was not governed by the clever men, " but by the active and energetic." He used also to quote another saying, that " common

" sense would be improving, when men left
" off wearing as much flour on their heads
" as would make a pudding ; when women
" left off wearing rings in their ears, like
" savages wear nose rings ; and when fire-
" grates were no longer made of polished
" steel."

Dr. Darwin has been frequently called an
atheist, whereas in every one of his works
distinct expressions may be found showing
that he fully believed in God as the Creator
of the universe. For instance, in the 'Temple
of Nature,' published posthumously,* he
writes: " Perhaps all the productions of
" nature are in their progress to greater per-
" fection ! an idea countenanced by modern
" discoveries and deductions concerning the
" progressive formation of the solid parts of
" the terraqueous globe, and consonant to
" the dignity of the creator of all things."
He concludes one chapter in ' Zoonomia '
with the words of the Psalmist: " *The*
" *heavens declare the Glory of God, and the*
" *firmament sheweth his handiwork.*"

* 'Temple of Nature,' 1803, note, p. 54. See also the
striking foot-note (p. 142) on the immutable properties of
matter " received from the hand of the Creator," etc.

He published an ode on the folly of atheism, with the motto " I am fearfully and wonderfully made," of which the first verse is as follows :—

'1.

Dull atheist, could a giddy dance
Of atoms lawless hurl'd
Construct so wonderful, so wise,
So harmonised a world ?

With reference to morality he says : * " The " famous sentence of Socrates, ' Know your- " self,' . . . . however wise it may be, seems " to be rather of a selfish nature. . . . . . " But the sacred maxims of the author of " Christianity, ' Do as you would be done by,' " and ' Love your neighbour as yourself,' " include all our duties of benevolence and " morality ; and, if sincerely obeyed by all " nations, would a thousandfold multiply the " present happiness of mankind."

Although Dr. Darwin was certainly a theist in the ordinary acceptation of the term, he disbelieved in any revelation. Nor did he feel much respect for unitarianism, for he used to say that " unitarianism was

* ' Temple of Nature,' 1803, note p. 124.

a feather-bed to catch a falling Christian."

Remembering through what an exciting period of history Erasmus lived, it is singular how rarely there is more than an allusion in his letters to politics. He would now be called a liberal, or perhaps rather a radical. He seems to have wished for the success of the North American colonists in their war for independence; for he writes to Wedgwood (Oct. 17, 1782): "I hope Dr. Franklin will " live to see peace, to see America recline " under her own vine and fig-tree, turning " her swords into plough-shares, &c." Like so many other persons, he hailed the beginning of the French Revolution with joy and triumph. Miss Seward, in a letter to Dr. Whalley, dated May 18, 1792, says: "I " should indeed now begin to fear for France; " but Darwin yet asserts that, in spite of all " disasters, the cause of freedom will triumph, " and France become, ere long, an example, " prosperous as great, to the surrounding " nations."

She remarks in another letter, Darwin " was

‘ a far-sighted politician, and foresaw and
‘“ foretold the individual and ultimate mis-
“ chief of every pernicious measure of the
“ late Cabinet.” *

In February 1789, he tells Wedgwood that
he had been reading ‘ Colonel Jack,’ by De
Foe, and suggests that the account there
given of the generous spirit of black slaves
should be republished in some journal.
Again, on April 13th of the same year
(1789), he writes: “I have just heard that
“ there are muzzles or gags made at Birming-
“ ham for the slaves in our islands. If this
“ be true, and such an instrument could be
“ exhibited by a speaker in the House of
“ Commons, it might have a great effect.
“ Could not one of their long whips or
“ wire tails be also procured and exhibited ?
“ But an instrument of torture of our own
“ manufacture would have a greater effect,
“ I dare say.”

The following lines on Slavery were
published in Canto III. of the ‘ Loves of
the Plants,’ 1790 :—

* ‘Journals of Dr. Whalley, 1863, vol. ii. pp. 73, 220–222.

> " Throned in the vaulted heart, his dread resort,
>   Inexorable Conscience holds his court ;
>   With still small voice the plots of Guilt alarms.
>   Bares his mask'd brow, his lifted hand disarms;
>   But wrapp'd in might with terrors all his own,
>   He speaks in thunder, when the deed is done.
>   Hear him, ye Senates ! hear this truth sublime,
>   He, who allows oppression, shares the crime."

The date of this poem and of the above
letter should be noticed, for let it be remembered that even the slave-trade was
not abolished until 1807 ; and in 1783 the
managers of the Society for the Propagation
of the Gospel absolutely declined, after a
full discussion, to give Christian instruction
to their slaves in Barbadoes. *

He sympathised warmly with Howard's
noble work of reforming the state of the
prisons throughout Europe, as his lines in
the ' Loves of the Plants ' (Canto II.)
show :—

> " And now, Philanthropy ! thy rays divine
>   Dart round the globe from Zembla to the line ;
>   O'er each dark prison plays the cheering light,
>   Like northern lustres o'er the vault of night.—
>   From realm to realm, with cross or crescent crown'd,
>   Where'er mankind and misery are found,

* Lecky, ' Hist. of England in the Eighteenth Century,' 1878,
vol. ii. p. 17.

O'er burning sands, deep waves, or wilds of snow,
Thy Howard journeying seeks the house of woe—
Down many a winding step to dungeons dank,
Where anguish wails aloud, and fetters clank;
To caves bestrew'd with many a mouldering bone,
And cells, whose echoes only learn to groan;
Where no kind bars a whispering friend disclose,
No sunbeam enters, and no zephyr blows,
He treads, inemulous of fame or wealth,
Profuse of toil, and prodigal of health;
With soft assuasive eloquence expands
Power's rigid heart, and opes his clenching hands,
Leads stern-eyed Justice to the dark domains,
If not to sever, to relax the chains.
  The spirits of the Good, who bend from high
Wide o'er these earthly scenes their partial eye,
When first, arrayed in Virtue's purest robe,
They saw her Howard traversing the globe;
Mistook a mortal for an Angel-Guest,
And ask'd what Seraph-foot the earth imprest.
Onward he moves! Disease and Death retire,
And murmuring demons hate him, and admire."

Judging from his published works, letters,
and all that I have been able to gather about
him, the vividness of his imagination seems
to have been one of his pre-eminent characteristics. This led to his great originality of
thought, his prophetic spirit both in science
and in the mechanical arts, and to his over-
powering tendency to theorise and generalise.

Nevertheless, his remarks, hereafter to be given, on the value of experiments and the use of hypotheses show that he had the true spirit of a philosopher. That he possessed uncommon powers of observation must be admitted. The diversity of the subjects to which he attended is surprising. But of all his characteristics, the incessant activity or energy of his mind was, perhaps, the most remarkable. Mr. Keir, himself a distinguished man, who had seen much of the world, and who " had been well acquainted " with Dr. Darwin for nearly half a century," after his death wrote (May 12th, 1802) to my father : " Your father did indeed retain more " of his original character than almost any " man I have known, excepting, perhaps, Mr. " Day [author of 'Sandford and Merton,' &c.]. " Indeed, the originality of character in both " these men was too strong to give way to " the example of others." He afterwards proceeds : " Your father paid little regard to "authority, and he quickly perceived the " analogies on which a new theory could be " founded. This penetration or sagacity by " which he was able to discover very remote

E

" causes and distant effects, was the charac-
" teristic of his understanding. Perhaps it
" may be thought in some instances to have
" led him to refine too much, as it is difficult
" in using a very sharp-pointed instrument
" to avoid sometimes going rather too deep.
" By this penetrating faculty he was enabled
" not only to trace the least conspicuous
" indications of scientific analogy, but also
" the most delicate and fugitive beauties of
" poetic diction. If to this quality you add
" an uncommon activity of mind and facility
" of exertion, which required the constant
" exercise of some curious investigation, you
" will have, I believe, his principal features."

His activity continued to his latest days;
and the following letter, written when he
was sixty-one years old to my father, shows
his continued zeal in his profession.

ERASMUS DARWIN *to* HIS SON ROBERT.

DEAR ROBERT,   DERBY, *April* 13, 1792.

I think you and I should sometimes exchange
a long medical letter, especially when any uncommon
diseases occur; both as it improves one in writing

clear intelligible English, and preserves instructive cases. Sir Joshua Reynolds, in one of his lectures on pictorial taste, advised painters, even to extreme old age, to study the works of all other artists, both ancient and modern; which he says will improve their invention, as they will catch collateral ideas (as it were) from the pictures of others, which is a different thing from imitation; and adds, that if they do not copy others, they will be liable to copy *themselves*, and introduce into their work the same faces, and the same attitudes again and again. Now in medicine I am sure unless one reads the work of others, one is liable perpetually to copy one's *own* prescriptions, and methods of treatment; till one's whole practice is but an imitation of one's self; and half a score medicines make up one's whole materia medica; and the apothecaries say the doctor has but 4 or 6 prescriptions to cure all diseases.

Reasoning thus, I am determined to read all the new medical journals which come out, and other medical publications, which are not too voluminous; by which one knows what others are doing in the medical world, and can astonish apothecaries and surgeons with the new and wonderful discoveries of the times. All this harangue lately occurred to me on reading the trials made by Dr. Crawford.

*   *   *   *   *

My father seems to have urged him, about the year 1793, to leave off professional work;

he answered, "it is a dangerous experiment,
" and generally ends either in drunkenness
" or hypochrondriacism. Thus I reason,
" one must do something (so country squires
" fox-hunt), otherwise one grows weary of
" life, and becomes a prey to ennui. There-
" fore one may as well do something advan-
" tageous to oneself and friends or to mankind,
" as employ oneself in cards or other things
" equally insignificant." During his frequent
and long journeys, he read and wrote much
in his carriage, which was fitted up for
the purpose. Nor was travelling an easy
affair in those days, for owing to the state of
the roads, a carriage could hardly reach some
of the houses which he had to visit; and I hear
from one of his granddaughters that an old
horse named the " Doctor," with a saddle on,
used to follow behind the carriage, without
being in any way fastened to it; and when
the road was too bad, he got out and rode
upon Doctor. This horse lived to a great age,
and was buried at the Priory.

When at home he was an early riser; and
he had his papers so arranged (as I have
heard from my father) that if he awoke in

the night he was able to get up and continue
his work for a time, until he felt sleepy.
Considering his indomitable activity, it is a
singular fact that he suffered much from a
sense of fatigue. On my once remarking to
my father, how greatly fatigued he seemed
to be after his day's work, he answered, " I
" inherit it from my father."

In some notes made by my father in 1802,
he states that Dr. Darwin naturally was of a
bold disposition, but that a succession of accidents made a deep impression on his mind, and
that he became very cautious. When he was
about five years old he received an accidental
blow on the top of his head, sufficiently severe
to give him a white lock of hair for life.
Later on, when he was fishing with his
brothers, they put him into a bag with only
his feet out, and being thus blinded he walked
into the river, and was very nearly drowned.
Again, when he and Lord George Cavendish
were playing with gunpowder at school, it
exploded, and he was badly injured; and
lastly, he broke his kneecap.

Owing to his lameness, he was clumsy in
his movements, but when young, was a very

active man.   His frame was large and bulky,
and he grew corpulent when old.   He was
deeply pitted with the small-pox.

It is remarkable that in so large a town
as Derby, and at so late a period as 1784,
there was no public institution for the
relief of the poor in sickness.   Dr. Darwin
therefore at this time drew up a circular, the
MS. of which is in my possession, stating that
" as the small-pox has already made great
" ravages in Derby, showing much malignity
" even at its commencement; and as it is now
" three years since it was last epidemic in this
" town, there is great reason to fear that it will
" become very fatal in the approaching spring,
" particularly amongst the poor, who want
" both the knowledge and the assistance neces-
" sary for the preservation of their children."
He accordingly proposed that a society should
be formed—the members to subscribe a guinea
each, and that a room should be hired as a
dispensary, where the medical men of the town
might give their attendance gratuitously.
The poor were to be directed to take their prescriptions in due order to all the druggists in
the town, apparently to disarm opposition.

The circular then expresses the hope that
the dispensary "may prove the foundation-
" stone of a future infirmary."

In this same year of 1784 he seems to
have taken the chief part in founding a
Philosophical Society in Derby. The members met for the first time at his house, and
he delivered to them a short but striking
address, from which the following passages
may be given: "I come now to the second
" source of our accurate ideas. As we are
" fashioned and constituted by the niggard
" hand of Nature with such imperfect and
" contracted faculties, with so few and such
" imperfect senses; while the bodies, which
" surround us, are indued with infinite variety
" of properties; with attractions, repulsions,
" gravitations, exhalations, polarities, minute-
" ness, irresistance, &c., which are not cog-
" nizable by our dull organs of sense, or not
" adapted to them; what are we to do? shall
" we sit down contented with ignorance, and
" after we have procured our food, sleep away
" our time like the inhabitants of the woods
" and pastures?   No, certainly!—since there
" is another way by which we may indirectly

" become acquainted with those properties of
" bodies, which escape our senses; and that
" is by *observing and registering their effects upon*
" *each other.* This is the tree of knowledge,
" whose fruit forbidden to the brute creation
" has been plucked by the daring hand of
" *experimental philosophy.*"

He concludes the address with the words:
" I hope at some distant time, perhaps not
" very distant, by our own publications we
" may add something to the common heap
" of knowledge; which I prophecy will never
" cease to accumulate, so long as the human
" footstep is seen upon the earth."

No man has ever inculcated more persist-
ently and strongly the evil effects of intemper-
ance than did Dr. Darwin; but chiefly on the
grounds of ill-health, with its inherited consequences; and this perhaps is the most practical
line of attack. It is positively asserted that he
diminished to a sensible extent the practice of
drinking amongst the gentry of the county.*

---

* The following short history of temperance societies is
extracted from Dr. Krause's MS. notes on Dr. Darwin :—
" The oldest temperance societies were founded in North
America in 1808 by the efforts of Dr. Rush, and in Great
Britain in 1829, chiefly at the suggestion of Mr. Dunlop.

He himself during many years never touched
alcohol under any form ; but he was not a
bigot on the subject, for in old age he informed
my father that he had taken to drink daily
two glasses of home-made wine with advantage. Why he chose home-made wine is not
obvious ; perhaps he fancied that he thus did
not depart so widely from his long-continued
rule. He also wrote (Oct. 15, 1772) to
Wedgwood, who had feeble health : " I would
" advise you to live as high as your constitution
" will admit of, in respect to both eating and
" drinking. This advice could be given to very
" few people ! If you were to weigh yourself
" once a month you would in a few months
" learn whether this method was of service to
" you." His advocacy of the cause is not yet

See Samuel Couling, 'History of the Temperance Movement in Great Britain and Ireland, from the earliest date
to the present time ;' London, 1862. In Germany, indeed, the
Archduke Frederick of Austria had founded a temperance order
as early as 1439, which was followed in 1600 by the temperance
order established by the Landgrave of Hesse, but these were only
imitations of the Templars and other orders of knighthood,
which sought by vows to suppress the coarse excesses of drinking
bouts, as is indicated by the motto of the first-mentioned order :
' Halt Maas ! ' The suggestion of the establishment in Germany of true temperance societies on the American and English
model was due to King Frederick William III."

forgotten, for Dr. Richardson, in his address
in 1879 to the " British Medical Temperance
Association," remarks: " the illustrious Hal-
" ler, Boerhaave, Armstrong, and particularly
" Erasmus Darwin, were earnest in their sup-
" port of what we now call the principles of
" temperance."

When a young man he was not always
temperate. Miss Seward relates* a story,
which would not have been worth notice had
it not been frequently quoted. My grandfather went on a picnic party in Mr. Sneyd's
boat down the Trent, and after luncheon, when
(in Miss Seward's elegant language), " if not
" absolutely intoxicated, his spirits were in a
" high state of vinous exhilaration," he suddenly got out of the boat, swam ashore in his
clothes, and " walked coolly over the meadows
" towards the town" of Nottingham. He
there met an apothecary, whose remonstrances
about his wet clothes he answered by saying
that the unusual internal stimulus would
" counteract the external cold and moisture;"
he then mounted on a tub, and harangued
the mob in an extremely sensible manner on

* 'Memoirs of Dr. Darwin,' pp. 64–68.

sanitary arrangements. But it is obvious
that these harangues must have been largely
the work of Miss Seward's own imagination.
There was, however, some truth in this story,
for his widow, who did not believe a word
of it, wrote to Mr. Sneyd, whose answer
lies before me. He admits that something
" similar" did happen, but gives no details,
and advises Mrs. Darwin " to take no notice
" of this part of her (Miss Seward's) very
" unguarded and scandalous publication."
To show what the gentry of the county
thought of her book at the time, I will add
that Mr. Sneyd, after alluding in the same
letter to her account of the death of his son
Erasmus, remarks : " The authoress deserves
" to be exposed for her want of veracity and
" every humane feeling." One of Dr. Darwin's stepsons (as I hear from his daughter)
used always to maintain that this half-tipsy
freak was due to some of the gentlemen of
the party, " who were vexed at his tempe-
" rate habits," having played him a trick;
and this, I presume, means that he was persuaded to drink something as weak which
was really strong.

The following incident related by Mr. Edgeworth * illustrates the humane side of his character. Mr. Edgeworth had corresponded, as a stranger, with Dr. Darwin, about the construction of carriages, and came to Lichfield to see him, but did not find him at home. He was asked by Mrs. Darwin to stay to supper. " When this was nearly " finished, a loud rapping at the door an- " nounced the doctor. There was a bustle " in the hall, which made Mrs. Darwin get " up and go to the door. Upon her ex- " claiming that they were bringing in a dead " man, I went to the hall. I saw some per- " sons, directed by one whom I guessed to be " Doctor Darwin, carrying a man who ap- " peared motionless. ' He is not dead,' said " Dr. Darwin, ' he is only dead drunk. I " found him,' continued the doctor, ' nearly " suffocated in a ditch; I had him lifted into " my carriage, and brought hither, that we " might take care of him to-night.' " Not many men would have done anything so disagreeable as to bring home a drunken man in their carriage. When a light was

* 'Memoirs of R. L. Edgeworth,' 2nd edit. vol. i. p. 158.

brought, the man was found to be, to the
astonishment of all present, Mrs. Darwin's
brother, "who for the first time in his life,"
as Mr. Edgeworth was assured, "had been
" intoxicated in this manner, and who would
" undoubtedly have perished had it not been
" for Dr. Darwin's humanity." We must
remember that in those good old days it was
not thought much of a disgrace to be very
drunk. After the man had been put to bed,
Mr. Edgeworth says that Dr. Darwin and he
first discussed the construction of carriages
and then various literary and scientific subjects, so that "he discovered that I had
" received the education of a gentleman."
"Why, I thought," said the doctor, "that
" you were only a coachmaker." "That was
" the reason," said I, "that you looked so sur-
" prised at finding me at supper with Mrs.
" Darwin. But you see, doctor, how superior
" in discernment ladies are even to the most
" learned gentleman."

He was kind and considerate to his servants,
as the two following stories show. His son
Robert owed him a small sum of money, and
instead of being paid, he asked Robert to buy

a goose-pie with it, for which it seems Shrewsbury was then famous, and send it at Christ-
mas to an old woman living in Birmingham,
" for she, as you may remember, was your
" nurse, which is the greatest obligation, if
" well performed, that can be received
" from an inferior." This was in the year
1793.

On the day of his death, in the early morning, whilst writing a long and affectionate
letter to Mr. Edgeworth, he was seized with a
violent shivering fit, and went into the kitchen
to warm himself before the fire.   He there saw
an old and faithful maid servant churning,
and asked her why she did this on a Sunday
morning.   She answered that she had always
done so, as he liked to have fresh butter
every morning.   He said : " Yes, I do, but
" never again churn on a Sunday !"

That Dr. Darwin was charitable, we may
believe on Miss Seward's testimony, as it is
supported by concurrent evidence.   After
saying that he would not take fees from the
priests and lay-vicars of the Cathedral of Lichfield, she adds : " Diligently, also, did he
" attend to the health of the poor in the city

" and afterwards at Derby, and supplied their
" necessities of food, and all sort of charitable
" assistance." *   Sir Brooke Boothby also, in
one of his published sonnets, says :—

> If bright example more than precept sway
> Go, take your lesson from the life of Day,
> Or, Darwin, thine whose ever-open door
> Draws, like Bethesda's pool, the suffering poor
> Where some fit cure the wretched all obtain
> Relieved at once from poverty and pain.

The gratitude of the poor to him was shown
on two occasions in a strange manner.†
Having to see a patient—one of the Caven-
dishes—at Newmarket during the races, he
slept at an hotel, and during the night was
awakened by the door being gently opened.
A man came to his bedside and thus spoke

* 'Memoirs of the Life of Dr. Darwin,' 1804, p. 5.

† These stories appear at first hardly credible, but I have
traced them, more or less clearly, through four distinct channels
to my grandfather, whose veracity has never been doubted by
any one who knew him.   The fundamental facts are the same
with respect to the jocky story, but the accessories differ to an
extreme degree.   With respect to the second story even some of
the fundamental facts differ, and I feel much doubt about it.   It
is quite curious how stories get unintentionally altered in the
course of years.   They were first communicated to me by a
daughter of Violetta Darwin, who heard her mother relate
them.

to him: "I heard that you were here, but
" durst not come to speak to you during the
" day. I have never forgotten your kind-
" ness to my mother in her bad illness, but
" have not been able to show you my grati-
" tude before. I now tell you to bet largely
" on a certain horse (naming one), and not
" on the favourite, whom I am to ride, and
" who we have settled is not to win." My
grandfather afterwards saw in the newspaper
that to the astonishment of everyone, the
favourite had not won the race.

The second story is, that as the doctor was
riding at night on the road to Nottingham a
man on horseback passed him, to whom he
said good night. As the man soon slackened
his pace, Dr. Darwin was forced to pass him,
and again spoke, but neither time did the man
give any answer. A few nights afterwards
a traveller was robbed at nearly the same
spot by a man who, from the description, appeared to be the same. It is added that my
grandfather out of curiosity visited the robber
in prison, who owned that he had intended
to rob him, but added: " I thought it was you,
" and when you spoke I was sure of it. You

" saved my life many years ago, and nothing
" could make me rob you."*

Notwithstanding so much evidence of Dr.
Darwin's benevolence and generosity, it has
been represented that he valued money in-
ordinately, and that he wrote only for gain.
This is the language of a notice published
shortly after his death,† which also says that
he was very vain, and that "flattery was
" found to be the most successful means of
" gaining his notice and favour."

All that I have been able to learn goes to
show that this was a mistaken view of his
character.

In a letter to my father, dated Feb. 7, 1792,
he writes:

" As to fees, if your business pays you well
" on the whole, I would not be uneasy about
" making absolutely the most of it. To live
" comfortably all one's life, is better than to
" make a very large fortune towards the end
" of it."

In another letter not dated, but written in

---

* In one version of this story, the visit of Dr. Darwin to the
man is not mentioned, so that there is then no point to the story.

† 'Monthly Magazine,' or 'British Register,' vol. xiii. 1802,
p. 457.

1793, he remarks: "There are two kinds of
" covetousness, one the fear of poverty, the
" other the desire of gain. The former, I
" believe, at some times affects all people
" who live by a profession." Again, his son
Erasmus, in writing on Nov. 12, 1792, to my
father, after remarking how rich he was becoming, adds: " I am not afraid of being rich,
" as our father used to say at Lichfield he
" was, for fear of growing covetous; to avoid
" which misfortune, as you know, he used to
" dig a certain number of duck puddles
" every spring, that he might fill them up
" again in the autumn." How it was possible
to expend much money in digging duck
puddles, it is not easy to see.

It is probable that the only foundation for
the reviewer's statements, and for others of a
like kind, was the habit he had—perhaps a
foolish one—of often speaking about himself
in a quizzing or bantering tone. Mr. Edgeworth, who had known him "intimately
" during thirty-six years," in answer to the
reviewer, writes :*

" I am most anxious to contradict that

* 'Monthly Magazine, vol. ii. 1802, p. 115.

" assertion of the anonymous biographer,
" which I consider the most unfounded and
" injurious—that Dr. Darwin wrote chiefly
" for money. . . . It is not improbable
" that to avoid offensive adulation he might
" have said ironically that his object in
" writing was money, not fame. I have
" heard him say so twenty times, but I never
" for one moment supposed him to be in
" earnest. . . . It is asserted by the re-
" viewer ' that he stooped to accept of gross
" 'flattery.' Perhaps in the inmost recesses
" of his heart, vanity might reign without
" control, but no man exacted less tribute of
" applause in conversation. When the admi-
" rable *travestie* of his poetic style was pub-
" lished in the Anti-jacobin newspaper, I
" spoke of it in his presence in terms of
" strong approbation, and he appeared to
" think as I did, of the wit, ingenuity, and
" poetic merit of the parody." To ask the
author of the ' Loves of the Plants ' to admire
the ' Loves of the Triangles ' was putting his
temper through a severe ordeal. Mr. Keir,
who had known Dr. Darwin well for nearly
half a century, remarks in a letter (May 12,

1802): "The works of your father are a
"more faithful monument and more true
"mirror of his mind than can be said of
"those of most authors. For he was not
"one of those who wrote *invitâ Minervâ*, or
"from any other incitement than the ardent
"love of the subject."

Throughout his letters I have been struck
with his indifference to fame, and the complete absence of all signs of any over-estima-
tion of his own abilities or of the success of
his works. I infer, from his having mentioned the fact to my father, that he was
pleased by receiving a print of himself, "well
"done, I believe—proofs, 10s. 6d.—the first
"impression of which the engraver, Mr.
"Smith, believes will soon be sold, and he
"will then sell a second at 5s." He then
adds: "but the great honour of all is to have
"one's head upon a sign-post, unless, indeed,
"upon Temple Bar!" This engraving was
copied from the picture by Wright of Derby,
of which a photograph is given in the present volume. Many pictures were made of
him, but with one or two exceptions they are
characterised by a rather morose and discon-

tented expression. Mr. Edgeworth, in writing
to him about one of these pictures, says :
" There is a cloud over your brow and a
" compression of the lips that hide your
" benevolence and good humour. And great
" author as you are, dear doctor, I think
" you excel the generality of mankind as
" much in generosity as in abilities."*

I have said that, as far as I can judge, he
was remarkably free from vanity, conceit, or
display ; nor does he appear to have been am-
bitious for a higher position in society. Miss
Fielding, a granddaughter of Lady Charlotte
Finch, governess to Queen Charlotte's daughters, was taken to Dr. Darwin, at Derby, on
account of her health, and was invited to stay
some time at his house. George the Third
heard of my grandfather's fame through
Lady Charlotte, and said : " Why does not
" Dr. Darwin come to London ? He shall
" be my physician if he comes " ; and he
repeated this over and over again in his
usual manner. But Dr. Darwin and his wife
agreed that they disliked the thoughts of a
London life so much, that the hint was not

* 'Memoirs,' 2nd ed. vol. ii. p. 177.

acted on. Others have expressed surprise that
he never migrated to London.

That he was irascible there can be no doubt.
My father says " he was sometimes violent in
" his anger, but his sympathy and benevo-
" lence soon made him try to soothe or soften
" matters." Mr. Edgeworth also says :* " Five
" or six times in my life I have seen him
" angry, and have heard him express that
" anger with much real, and more apparent
" vehemence—more than men of less sensi-
" bility would feel or show. But then the
" motive never was personal. When Dr.
" Darwin beheld any example of inhumanity
" or injustice, he never could refrain his
" indignation ; he had not learnt, from the
" school of Lord Chesterfield, to smother
" every generous feeling."

In 1804 Miss Seward published her ' Life
of Dr. Darwin.' It was unfortunate for his
fame that she undertook this task, for she
knew nothing about science or medicine, and
the pretentiousness of her style is extremely
disagreeable, not to say nauseous, to many

* 'Monthly Magazine,' 1802, p. 115.

persons; though others like the book much.
It abounds with inaccuracies, as both my
father and other members of the family
asserted at the time of its publication. For
instance, she states that when dying he sent
for Mrs. Darwin, and first asked her and
then his daughter Emma to bleed him, and
gives their answers in inverted commas. But
the whole account is a simple fiction, for he
expressly told his servant not to call Mrs.
Darwin, but was disobeyed as the servant
saw how ill he was; and his daughter was
not even present. She does not even give
his age at the time of his death correctly. It
is also obvious that the many long speeches
inserted in her book are the work of her own
imagination, either with some or with no
foundation.

She describes (p. 406) his conduct when
he heard of the suicide of his son Erasmus,
who drowned himself during a fit of temporary insanity, as inhuman to an unpa-
ralleled degree. She asserts that when he
was told that the body " was found, he ex-
" claimed in a low voice, 'Poor insane
" 'coward,' and, it is said, never afterwards
" mentioned the subject." Miss Seward then

proceeds (p. 408), " this self-command enabled
" him to take immediate possession of the
" premises bequeathed to him (by his son
" Erasmus); to lay plans for their improve-
" ment; to take pleasure in describing those
" plans to his acquaintance, and to determine
" to make it his future residence ; and all this
" without seeming to recollect to how sad an
" event he owed their possession !"

The whole of this account is absolutely
false, and when my father demanded her
authority, she owned that it had been given
merely on a report at a distant place, without
any inquiry having been made from a single
person who could have really known what
happened. On the day after the death of
his son (Dec. 30th, 1799), in a letter to
my father, he says: " I write in great
" anguish of mind to acquaint you with a
" dreadful event—your poor brother Erasmus
" fell into the water last night at the bottom
" of his garden, and was drowned." His
daughter Emma, who was with him when
the news was brought to him that the body
had been at last found, gave the following
account of his behaviour to my mother : " He
" immediately got up, but staggered so much

" that Violetta and I begged of him to sit
" down, which he did, and leaned his head
" upon his hand . . . . he was exceedingly
" agitated, and did not speak for many
" minutes. His first words were, ' I beg you
" ' will not, any of you, ask to see your poor
" ' brother's corpse;' and upon our assuring
" him that we had not the least wish to do
" so, he soon after said that this was the
" greatest shock he had felt since the death of
" his poor Charles." Emma then asserts that
Miss Seward's other statements are utterly
false, namely, that he never afterwards mentioned his son's death, and that he took immediate possession of the property bequeathed to
him. After alluding to other inaccuracies, in
Miss Seward's book, Emma concludes in a
truly feminine and filial spirit: " There is
" nothing else of such infinite consequence as
" her daring publicly to accuse my dear papa of
" want of affection and feeling towards his son.
" How can this be contradicted ? I want to
" scratch a pen over all the lies, and send the
" book back to Miss Seward; but mamma
" won't allow this. She thinks you and my
" brother will think of a better plan; for
" myself, I should feel no objection to swear

" the truth of what I have said before both
" houses of Parliament."

In one of my grandfather's letters, dated
Feb. 8th, 1800, he writes: "I am obliged as
" executor daily to study his (Erasmus's) ac-
" counts, which is both a laborious and
" painful business to me." A fortnight afterwards he tells my father about a monument
to be erected to Erasmus, and adds: "Mrs.
" Darwin and I intend to lie in Breadsall
" church by his side." Rarely has a more unfounded calumny been published about anyone
than the above account given by Miss Seward
of Dr. Darwin's behaviour when he heard of
his son's death.*

That the act of suicide was committed during
temporary insanity there can be little doubt. It

* Miss Seward published, on my father's demand, the following retractation in several journals, but such retractations are
soon forgotten, and the stigma remains: "The authoress of the
'Memoirs of Dr. Darwin,' since they were published, has discovered, on the attestation of his family and other persons present at the juncture, that the statement given of his exclamation,
page 406, on the death of Mr. Erasmus Darwin, is entirely without foundation; and that the doctor, on that melancholy event,
gave amongst his own family, proofs of strong sensibility at the
time, and of succeeding regard to the memory of his son, which
he seemed to have a pride in concealing from the world. In
justice to his memory, she is desirous to correct the misinforma-
tion she had received." ('Monthly Magazine,' 1804, p. 378;
and other journals and newspapers.)

is known that a change of disposition generally
precedes insanity, and Erasmus, from being
an excellent man of business, had become
dilatory to an abnormal degree. It appears
that he had neglected to do something of importance for my father; and my grandfather,
nearly two years before Erasmus's death,
wrote in his excuse to my father (Jan. 8th,
1798) as follows: "I have not spoken to
" him on your affairs, his neglect of small busi-
" nesses (as he thinks them, I suppose,) is a
" constitutional disease. I learnt yesterday that
" he had like to have been arrested for a small
" candle bill of 3 or 4 pounds in London, which
" had been due 4 or 5 years, and they had
" repeatedly written to him! and that a trades-
" man in this town has repeatedly complained
" to a friend of his that he owes Mr. D.
" £70, and cannot get him to settle his ac-
" count. I write all this to show you, that his
" neglectful behaviour to you, was not owing
" to any disrespect, or anger, but from what?
" —*from defect of voluntary power.* Whence
" he procrastinates for ever!"

He was evidently conscious himself of some
mental change, for he purchased, six weeks

before his death, the small estate of the
Priory, near Derby; where he intended,
though only forty years old, to retire from
business, and spend the rest of his days in
quiet; or, as Dr. Darwin, who could not
have foreseen what all this foreboded, expressed it (in a letter to my father, Nov.
28th, 1799), "to sleep away the remainder of
" his life."

Amongst the property of Erasmus my
grandfather found a little cross made of platted
grass (now in my possession) gathered from
the tomb of Charles, who had died twenty
years before. A week before his own death,
he sent this to my father to be preserved.

The false reports about Dr. Darwin's
conduct on the death of his son, probably
originated in his strong dislike to affectation,
or to any display of emotion in a man. He
therefore wished to conceal his own feelings,
and perhaps did so too effectually. My father
writes : " He never would allow any common
" acquaintance to converse with him upon
" any subject that he felt poignantly. . . . .
" It was his maxim, that in order to feel
" cheerful you must appear to be so." There

was, moreover, a vein of reserve in him. Miss
Seward, in answer to a remark by my father,
says (May 10th, 1802, i.e., before the publication of the ' Memoirs ') : " Too well was I
" acquainted with the disposition and habits
" of your lamented father, to feel surprise
" from your telling me how little you had
" been able to gather from himself concerning
" the circumstances of his life, which pre-
" ceded your birth, and those which passed
" beneath the unobservant eyes of sportive
" infancy."

The many friends and admirers of Dr. Darwin were indignant at Miss Seward's book, and
thought that it showed much malice towards
him. No such impression was left on my
mind when lately re-reading it, but only that
of scandalous negligence, together, perhaps,
with a wish to excite attention to her book,
by inserting any wild and injurious report
about him. The friends, however, of Dr.
Darwin were right, for in a letter, dated May
12th, 1802, written to the Rev. Dr. Whalley,*
before she published the Memoirs, she shows

* 'Journals of Dr. Whalley,' edited by Wickham; not published until 1863, vol. i. p. 342.

her true colours, and gives an odious character of "that large mass of genius and sarcasm," as she calls him: She speaks of the "cold " satiric atmosphere around him, repulsing " the confidence and the sympathy of friend- " ship." And adds in her usual stilted phrase, " Age did not improve his heart; and on its " inherent frost, poetic authorism, commenc- " ing with him after middle life, engrafted " all its irritability, disingenuous arts, and " grudging jealousy of others' reputation in " that science."

It is natural to inquire why Miss Seward wrote so bitterly about a man with whom she had lived on intimate terms during many years, and for whom she often expressed, and probably felt, the highest admiration. The only possible explanation appears to be that she had wished to marry him after the death of his first wife and before his second marriage. This was the case according to several members of the family, and I under- stood from my father that he possessed docu- mentary evidence (subsequently destroyed) to this effect. This explains the following sig- nificant sentence in a letter written to her by

my father, March 5th, 1804, in relation to her
account of the suicide of Erasmus : " Were
" I to have published my father's papers in
" illustration of his conduct, some circum-
" stances must unavoidably have appeared,
" which would have been as unpleasant for
" you to read as for me to publish." Disap-
pointed affection, with some desire for
revenge, renders her whole course of conduct
intelligible.

I may here allude to some calumnies about
Dr. Darwin, which appeared in 1858 in the
' Life' of Mrs. Schimmelpenninck, who was a
younger sister of Tertius Galton, Dr. Darwin's
son-in-law. She there says that he scoffed at
conscience and morality, disbelieved in God,
and was a coarse glutton. These statements
are hardly worth notice, as they were dictated
in old age, she having seen Dr. Darwin, in
her own words, only " with the eyes of a
child." Nor was she always a trustworthy
person. I have a copy of a letter written
(Feb. 20th, 1871) by one of her nieces to
Dr. Dowson, who had used her book in his
' Life of Dr. Darwin,' and nothing can be more
explicit than the remarks about her un-

trustworthiness.   One of her sisters also, in
speaking of these statements, says: " They
" are facts distorted, and give a false impres-
" sion."   With regard to the charge of glut-
tony, as Dr. Darwin was a tall, bulky man,
who lived much on milk, fruit, and vegetables,
it is probable that he ate largely, as every
man must do who works hard and lives on
such a diet.

As it is interesting to see how far Erasmus
Darwin transmitted his characteristic qualities
of mind to his descendants, I will give a short
account of his children.   He had three sons
by his first wife (besides two who died in
infancy), and four sons and three daughters
by his second wife.   His eldest son, Charles
(born September 3, 1758), was a young man
of extraordinary promise, but died (May 15,
1778) before he was twenty-one years old from
the effects of a wound received whilst dissect-
ing the brain of a child.   He inherited from
his father a strong taste for various branches of
science, for writing verses, and for mechanics.
"Tools were his playthings," and   making

" machines was one of the first efforts of his
" ingenuity, and one of the first sources of
" his amusement." *

He also inherited stammering. With the
hope of curing him, his father sent him to
France when about eight years old (1766–67),
with a private tutor, thinking that if he was
not allowed to speak English for a time, the
habit of stammering might be lost; and it is
a curious fact that in after years when speaking French he never stammered. At a very
early age he collected specimens of all kinds.
When sixteen years old he was sent for a year
to Oxford, but he did not like the place, and
" thought (in the words of his father) that the
" vigour of the mind languished in the pur-
" suit of classical elegance, like Hercules at
" the distaff, and sighed to be removed to the
" robuster exercise of the medical school of
" Edinburgh." He stayed three years at
Edinburgh, working hard at his medical

* These statements are taken chiefly from a sketch of his life
published by his father, Erasmus, in 1780, together with two of
his posthumous medical essays. See also Hutchinson's ' Bio-
graphia Medica,' 1799, vol. i. p. 239 ; also ' Biographie
Universelle,' vol. x. 1855; also an article in the ' Gentleman's
Magazine,' September 1st, 1794, vol. lxiv. p. 794, signed " A. D.,"
evidently Professor Andrew Duncan, of Edinburgh.

studies, and attending " with diligence all the
" sick poor of the parish of Waterleith, and
" supplying them with the necessary medi-
" cines." The Æsculapian Society awarded
him its first gold medal for an experimental
enquiry on pus and mucus. Notices of him
appeared in various journals; and all the
writers agree about his uncommon energy
and abilities. He seems, like his father, to
have excited the warm affection of his friends.
Professor Andrew Duncan, in whose family
vault Charles was buried, cut a lock of hair
from the corpse, and took it to a jeweller,
whose apprentice, afterwards the famous Sir
H. Raeburn, set it in a locket for a memorial.*
The venerable professor spoke to me about
him with the warmest affection forty-seven
years after his death, when I was a young
medical student in Edinburgh. The inscription on his tomb, written by his father, says,
with more truth than is usual on such occasions: " Possessed of uncommon abilities and
" activity, he had acquired knowledge in
" every department of medical and philoso-
" phical science, much beyond his years."

* ' Harveian Discourse,' by Professor A. Duncan, 1824.

Dr. Darwin was able to reach Edinburgh before Charles died, and had at first hopes of his recovery; but these hopes, as he informed my father, " with anguish," soon disappeared. Two days afterwards he wrote to Wedgwood to the same effect, ending his letter with the words, " God bless you, my dear friend, may " your children succeed better." Two and a half years afterwards he again wrote to Wedgwood, " I am rather in a situation to " demand than to administer consolation."

About the character of his second son, Erasmus (born 1759), I have little to say, for, though he wrote poetry, he seems to have had none of the other tastes of his father. He had, however, his own peculiar tastes, viz. genealogy, the collecting of coins, and statistics. When a boy he counted all the houses in the city of Lichfield, and found out the number of inhabitants in as many as he could; he thus made a census, and when a real one was first made, his estimate was found to be nearly accurate. His disposition was quiet and retiring. My father had a very high opinion of his abilities, and this was probably just, for he would not otherwise have been invited to travel with, and pay long visits to, men so distinguished in different ways as Boulton the engineer, and Day, the moralist and novelist. He was certainly very ingenious. He detected by a singularly subtle plan the author of a long series of anonymous letters, which had caused, during six or seven years, extreme annoyance and even misery to many of the inhabitants of the county. The author was found to be a county gentleman of not inconsiderable standing. He was a successful solicitor in Lichfield, but his death, Dec. 30, 1799, was a sad one, as I have already mentioned.

The third son, Robert Waring Darwin (my father, born 1766), did not inherit any aptitude for poetry or mechanics, nor did he possess, as I think, a scientific mind. He published, in Vol. lxxvi. of the 'Philosophical Transactions,' a paper on Ocular Spectra, which Wheatstone told me was a remarkable production for the period; but I believe that he was largely aided in writing it by his father. He was elected a Fellow of the Royal Society in 1788. I cannot tell why my father's mind did not appear to me fitted

for advancing science; for he was fond of
theorising, and was incomparably the most
acute observer whom I ever knew. But his
powers in this direction were exercised almost
wholly in the practice of medicine, and in the
observation of human character. He intui-
tively recognised the disposition or character,
and even read the thoughts, of those with
whom he came into contact with extraordinary acuteness. This skill partly accounts
for his great success as a physician, for it
impressed his patients with belief in him;
and my father used to say that the art of
gaining confidence was the chief element in
a doctor's worldly success.

Erasmus brought him to Shrewsbury
before he was twenty-one years old, and left
him £20, saying, "Let me know when you
"want more, and I will send it you." His
uncle, the rector of Elston, afterwards also
sent him £20, and this was the sole pecuniary
aid which he ever received. I have heard
him say that his practice during the first
year allowed him to keep two horses and a
man-servant. Erasmus tells Mr. Edgeworth
that his son Robert, after being settled in

Shrewsbury for only six months, "already
had between forty and fifty patients." By
the second year he was in considerable, and
ever afterwards in very large, practice. His
success was the more remarkable, as he for
some time detested the profession, and declared
that if he had been sure of gaining £100 a
year in any other way he would never have
practised as a doctor.

He had an extraordinary memory for the
dates of certain events, so that he knew the
day of the birth, marriage, and death of most
of the gentlemen of Shropshire. This power,
however, far from giving him any pleasure,
annoyed him, for he told me that his memory
for dates reminded him of painful events, and
so added to his regret for the death of old
friends. His spirits were generally high, and
he was a great talker. He was of an extremely sensitive nature, so that whatever
annoyed or pained him, did so to an extreme
degree. He was also somewhat easily roused
to anger. One of his golden rules was never
to become the friend of any one whom you
could not thoroughly respect, and I think he
always acted on it. But of all his charac-

teristic qualities, his sympathy was pre-eminent, and I believe it was this which made him for a time hate his profession, as it constantly brought suffering before his eyes. Sympathy with the joy of others is a much rarer endowment than sympathy with their pains, and it is no exaggeration to say that to give pleasure to others was to my father an intense pleasure. He died November 13th, 1849. A short notice of his life appeared in the 'Proceedings of the Royal Society.'

Of the children of Erasmus by his second marriage, one son became a cavalry officer, a second rector of Elston, and a third, Francis (born 1786, died 1859), a physician, who travelled far in countries rarely visited in those days. He showed his taste for Natural History by being fond of keeping a number of wild and curious animals. I may add that one of his sons, Captain Darwin, is a great sportsman, and has published a little book, the 'Gamekeeper's Manual' (4th ed. 1863), which shows keen observation and knowledge of the habits of various animals. The eldest daughter of Erasmus, Violetta, married S. Tertius Galton, and I feel sure that their son,

Francis,* will be willing to attribute the remarkable originality of his mind in large part
to inheritance from his maternal grandfather.†

   As Dr. Krause has so fully discussed Dr.
Darwin's published writings I have but little
to say about them.   After settling at Lichfield, he attended, during several years, chiefly
to medicine; but no doubt he was continually observing and making notes on various
subjects.   A huge folio common-place book,
begun in 1776, is in the possession of Reginald
Darwin and is half filled with notes and speculations.   Considering how voluminous a writer
he became when old, it is remarkable that he
does not appear to have thought for a long
time of publishing either prose or poetry.   In
a letter dated Nov. 21st, 1775, (ætat. 43) to

   * Author of " Hereditary Genius," " English Men of Science,"
and of other works and papers.
   † In the interval between his first and second marriages, Dr.
Darwin became the father of two illegitimate daughters.   In our
present state of society it may seem a strange fact that my
grandfather's practice as a physician should not have suffered by
his openly bringing up illegitimate children.   But to his credit
be it said that he gave them a good education, and from all that
I have heard they grew up to be excellent women, and lived on
intimate terms with his widow and the children by the second
marriage.

Mr. Cradock,* thanking him for a present of his 'Village Memoirs,' he says: " I have " for twenty years neglected the muses, and " cultivated medicine alone with all my in- " dustry . . . I lately interceded with " a Derbyshire lady to desist from lopping a " grove of trees, which has occasioned me to " try again the long-neglected art of verse- " making, which I shall inclose to amuse you, " promising, at the same time, never to write " another verse as long as I live, but to apply " my time to finishing a work on some branches " of medicine, which I intend for posthumous " publication."

In 1778 he purchased about eight acres of land near Lichfield, which he made into a botanic garden; and this seems to have been his chief amusement. " This wild umbrageous " valley . . . irriguous from various " springs, and swampy from their plenitude," as Miss Seward calls it,† now forms part of an adjoining park; and a Handbook for Lichfield describes it as still " a wild spot, but very " picturesque; many of the old trees remain-

---

* 'Literary Memoirs,' 1828, vol. iv. p. 143.
† 'Memoirs of the Life of Dr. Darwin,' 1804, p. 125.

" ing, and occasionally a few Darwinian snow-
" drops and daffodils peeping through the
" turf, and bravely fighting the battle of life."

This garden led him to write his poem of
the 'Botanic Garden,' the second part of
which, entitled the 'Loves of the Plants,' was
published, oddly enough, before the first part
called the 'Economy of Vegetation.' The
' Loves of the Plants,' judging from a prefixed
sonnet, must have appeared in 1788, and the
second edition in 1790. Miss Seward, in her
life of Dr. Darwin, accuses him of having
appropriated several of her verses, and of publishing them in this poem without any acknowledgment. The case is a very odd one; for
first, she herself admits* that it was entirely
through his instrumentality that these verses
were published with her name attached to
them, before the appearance of the ' Botanic
Garden,' in the ' Monthly Magazine,' and afterwards in the ' Annual Register.' Secondly,
there seems to have been little temptation for
the theft, for the whole history of his life
shows that writing verses on any subject was
not the least labour to him, but only a pleasure.

* ' Memoirs of the Life of Darwin,' p. 132.

And thirdly, that Miss Seward remained on the same friendly, almost playful, terms with him afterwards as before. The whole case is unintelligible, and in some respects looks more like highway robbery than simple plagiarism. Mr. Edgeworth, in a letter (Feb. 3, 1812) to Sir Walter Scott,* says that he had expressed surprise to Dr. Darwin at seeing Miss Seward's lines at the beginning of his poem, and that Dr. Darwin replied: " It was a compliment " which he thought himself bound to pay to a " lady, though the verses were not of the same " tenor as his own." But this seems a lame excuse, and it is an odd sort of compliment to take the verses without any acknowledgment. Perhaps he thought it fair play, for Edgeworth goes on to say that " Miss Seward's ' Ode " to Captain Cook ' stands deservedly high " in public opinion. Now to my certain know- " ledge most of the passages which have been " selected in the various reviews of the work " were written by Dr. Darwin. . . . I knew " him well, and it was as far from his temper " and habits, as it was unnecessary to his " acquirements, to beg, borrow, or steal from " any person on earth." These passages at

* ' Memoirs of R. L. Edgeworth,' 2nd ed. 1821, vol. ii. p. 245.

any rate show how true and ardent a friend
Edgeworth was to Dr. Darwin long after his
death.

In a letter to my father, dated Feb. 21st,
1788, he says: " I am printing the ' Loves of
" the Plants,' which I shall not put my name
" to, tho' it will be known to many. But
" the addition of my name would seem as if
" I thought it a work of consequence." Notwithstanding this depreciatory estimate, its
success was great and immediate ; and I have
heard my father, who was accurate about
figures, say that a thousand guineas were
paid before publication for the part which
was published last ; an amount which must
have been something extraordinary in those
days. Nor was the success quite transitory,
for a fourth edition appeared in 1799. In
1806 an octavo edition of all his poetical works
was published in three volumes. I have myself met with old men who spoke with a
degree of enthusiasm about his poetry, quite
incomprehensible at the present day. Horace
Walpole, in his letters repeatedly alludes
with admiration to Dr. Darwin's poetry, and
in a letter to Mr. Barrett (May 14th, 1792)
writes :—

" The ' Triumph of Flora,' beginning at the
" fifty-ninth line, is most beautifully and en-
" chantingly imagined ; and the twelve verses
" that by miracle describe and comprehend
" the creation of the universe out of chaos,
" are in my opinion the most sublime pas-
" sages in any author, or in any of the few
" languages with which I am acquainted.
" There are a thousand other verses most
" charming, or indeed all are so, crowded
" with most poetic imagery, gorgeous epi-
" thets and style : and yet these four cantos
" do not please me equally with the ' Loves of
" the Plants.' " The lines thus eulogised
are :—

" —— Let there be light ! " proclaimed the Almighty Lord.
Astonished Chaos heard the potent word ;—
Through all his realms the kindling Ether runs,
And the mass starts into a million suns ;
Earths round each sun with quick explosions burst,
And second planets issue from the first ;
Bend, as they journey with projectile force,
In bright ellipses their reluctant course ;
Orbs wheel in orbs, round centres centres roll,
And form, self-balanced, one revolving whole.
Onward they move amid their bright abode,
Space without bound, the Bosom of their God !

('The Botanic Garden,' part i. canto i. lines 103–114.)

Mr. Edgeworth, in a letter (1790) to
Dr. Darwin, writes about the 'Botanic Garden :'* "I may, however, without wounding
" your delicacy, say that it has silenced for
" ever the complaints of poets, who lament
" that Homer, Milton, Shakespeare, and a
" few classics, had left nothing new to de-
" scribe, and that elegant imitation of imita-
" tions was all that could be expected in
" modern poetry. . . . I read the descrip-
" tion of the Ballet of Medea to my sisters,
" and to eight or ten of my own family. It
" seized such hold of my imagination, that
" my blood thrilled back through my veins,
" and my hair broke the cementing of the
" friseur to gain the attitude of horror."
After the fame of his poetry had begun to
wane, Edgeworth predicted (p. 117) " that in
" future times some critic will arise who shall
" rediscover the 'Botanic Garden,' and build
" his fame upon the discovery." "It will
' shine out again, the admiration of posterity."

Several poets addressed him in compli-
mentary odes, as may be seen in the edition

* 'Memoir of R. L. Edgeworth,' 2nd ed. 1821, vol. ii.
p. 111.

of 1806. Cowper, who, one would have thought, differed in taste from him as much as one man could from another, yet, in conjunction with Hayley, wrote a poem in his honour,* beginning :

> No envy mingles with our praise,
>   Tho' could our hearts repine
> At any poet's happier lays,
>   They would, they must, at thine.

Notwithstanding the former high estimation of his poetry by men of all kinds in England, no one of the present generation reads, as it appears, a single line of it. So complete a reversal of judgment within a few years is a remarkable phenomenon. His verses were, however, quizzed by some persons not long after their publication. In the 'Pursuits of Literature,'† they are called :

> "Filmy, gauzy, gossamery lines.
> *   *   *   *   *
> Sweet tetrandrian, monogynian strains."

But the sudden downfall of his fame as a poet was in great part caused by the publication of the well-known parody the 'Loves of

* Dated June 23, 1793, and published in the Monthly Magazine,' 1803, vol. ii. p. 100.

† 'Pursuits of Literature.' A Satirical Poem in Four Dialogues; 14th ed. 1808, p. 54.

the Triangles.' No doubt public taste was
at this time changing, and becoming more
simple and natural. It was generally ac-
knowledged, under the guidance of Words-
worth and Coleridge, that poetry was chiefly
concerned with the feelings and deeper work-
ings of the mind ; whereas, Darwin maintained
that poetry ought chiefly to confine itself to
the word-painting of visible objects. He remarks ('Loves of the Plants': Interlude
between Cantos I. and II.) that poetry
should consist of words which express ideas
originally received by the organ of sight.
" . . . And as our ideas derived from visible
" objects are more distinct than those derived
" from the objects of our other senses, the
" words expressive of these ideas belonging
" to vision make up the principal part of
" poetic language. That is, the poet writes
" principally for the eye ; the prose writer
" uses more abstracted terms." It must be
admitted that he was a great master of language. In one of the earliest and best criti-
cisms on his poetry * it is said no man " had a

* 'Monthly Magazine or British Register,' 1802, vol. xiii.
pp. 457-463.

" more imperial command of words, or could
" elucidate with such accuracy and elegance
" the most complex and intricate machinery."
Byron called him " a mighty master of un-
meaning rhyme."

His first scientific publication was a paper
in the ' Philosophical Transactions' for 1757,
in which he confutes the view of Mr. Eeles,
that vapour ascends through " every particle,
being endued with a portion of electric fire."
The paper is of no value, but is curious as
showing in what a rudimentary condition
some branches of science then were. For Dr.
Darwin remarks that the " distinction has not
" been sufficiently considered by anyone to
" my knowledge " between " the immense
" rarefaction of explosive bodies " due " to the
" escape of air before condensed in them," as
when a few grains of gunpowder are exploded in a bladder, and to " the expansion
" of the constituent parts of those bodies "
through heat, as with steam.

The following speculative letter (though not
published) is interesting ; but its date must be
borne in mind in judging of its merits.

ERASMUS DARWIN *to* JOSIAH WEDGWOOD.

DEAR SIR,   *March*, 1784.

I admire the way in which you support your
new theory of freezing steam. You say, "Will not
vapour freeze with a less degree of cold than water
in the mass? instances hoar-frost, &c." Now this
same *et cætera*, my dear friend, seems to me to be a
gentleman of such consequence to your theory, that
I wish he would unfold himself a little more.

I sent an account of your experiment to Mr.
Robert, and desired him to show it to Dr. Black, so
that I shall hope some time to hear his opinion on
the very curious fact you mention, of a part of ice
(during a thaw) freezing whilst you applied a heated
body to another part of it. Now in spite of your
et cætera, I know no fact to ascertain that vapour
will freeze with less cold than water. I can in no
way understand why, during the time you apply a
heated body to one part of a piece of ice, when the
air of your room was at 50° and the ice had for a day
or two been in a thawing state, that a congelation
should be formed on another part of the same ice,
but from the following circumstances. There is
great analogy between the laws of the propagation of
heat, and those of electricity, such as the same bodies
communicate them easily, as metals, and the same
bodies with more difficulty, as glass, wax, air: they are
both excitable by friction, both give light, fuse metals,

et cætera. *Therefore I suppose that atmospheres of heat of different densities, like atmospheres of electricity, will repel each other at certain distances,* like globules of quicksilver pressed against each other, and that hence by applying a heated body near one end of a cold body, the more distant end may immediately become colder than the end nearest to the heated body.

March 11, 1784. Since I wrote the above I have reconsidered the matter, and am of opinion that steam, as it contains more of the element of heat than water, must require more absolute cold to turn it into ice, though the same sensible cold, as is necessary to freeze water, and that the phenomenon you have observed, depends on a circumstance which has not been attended to. When water is cooled down to freezing point, its particles come so near together, as to be within the sphere of their reciprocal attractions;—what then happens?—they accede with violence to each other and become a solid, at the same time pressing out from between them some air which is seen to form bubbles in ice and renders the whole mass lighter than water (on which it will swim) by this air having regained its elasticity; and pressing out any saline matters, as sea-salt, or blue vitriol, which have become dissolved in it; and lastly by thus forcibly acceding together, the particles of water press out also some more heat, as is seen by the rising of the thermometer immersed in such

freezing water. This last circumstance demands
your nice attention, as it explains the curious fact
you have observed. When the heat is so far taken
away from water, that the particles attract each
other, they run together with violence, and press out
some remaining heat, which existed in their interstices. Then the contrary must also take place when
you add heat to ice, so as to remove the particles
into their reciprocal spheres of repulsion : they recede
from each other violently, and thence attract more
heat into their interstices; and if your piece of hot
silver is become cold, and has no more heat to give,
or if this thawing water in this its expansile state is
in contact with other water which is saturated with
heat, it will rob it of a part, or produce freezing if
that water was but a little above 32°.

I don't know if I have expressed myself intelli-
gibly. I shall relate an experiment I made twenty-
five years ago, which confirms your fact. I filled a
brewing-copper, which held about a hogshead and
half, with snow; and immersed about half-an-ounce of
water at the bottom of a glass tube in this snow, as
near the centre as I could guess, and then making a
brisk and hasty wood-fire under it, and letting the
water run off by a cock as fast as it melted, I found
in a few minutes on taking out the tube that the water
in it was frozen. This experiment coincides with
yours, and I think can only be explained on the above
principle. In support of the above theory I can

prove from some experiments, that air when it is mechanically expanded always attracts heat from the bodies in its vicinity, and therefore water when expanded should do the same. But this would lengthen out my letter another sheet; I shall therefore defer it till I have the pleasure of a personal conference with you. Thus ice in freezing gives out heat suddenly, and in thawing gives out cold suddenly; but this last fact had not been observed (except in chemical mixtures) because when heat has been applied to thaw ice, it has been applied in too great quantities.

When shall we meet? Our little boy has got the ague, and will not take bark, and Mrs. Darwin is therefore unwilling to leave him, and begs to defer her journey to Etruria till later in the season. Pray come this way to London or from London. Our best compts. to all yours.

Adieu,

E. DARWIN.

P.S.—Water cooled beneath 32°, becomes instantly ice on any small agitation, or pouring out of one vessel into another, because that the accession of the particles to each other, and the pressing out of the air, or saline matters, and of heat is facilitated.

The 'Zoonomia,' which had been in preparation during many years, was published in

1794. We have seen that in 1775 it was intended for posthumous publication. Even so late as Feb. 1792, Dr. Darwin wrote to my father :—" I am studying my 'Zoonomia,' " which I *think* I shall publish, in hopes of " selling it, as I am now too old and hardened " to fear a little abuse. Every John Hunter " must expect a Jessy Foot to pursue him, as a " fly bites a horse." The work when published was translated into German, French, and Italian, and was honoured by the Pope by being placed in the 'Index Expurgatorius.' Dr. Krause has given so full, impartial, and interesting an account of the scientific views contained in this and his other works that I need say little on this head. Although he indulged largely in hypotheses, he knew full well the value of experiments. Maria Edgeworth, in writing (March 9th, 1792) about her little brother Henry, who was fond of collecting and observing, says :—" He will " at least never come under Dr. Darwin's " definition of a fool. ' A fool, Mr. Edgeworth, " ' you know, is a man who never tried an " ' experiment in his life. ' " * Again, in an

* 'Memoir of Maria Edgeworth,' 1867, vol. i. p. 31.

Apology, prefixed to the ' Botanic Garden,' we
have the following just remarks :—" It may
" be proper here to apologise for many of the
" subsequent conjectures on some articles of
" natural philosophy, as not being supported by
" accurate investigation, or conclusive experi-
" ments. Extravagant theories, however, in
" those parts of philosophy, where our know-
" ledge is yet imperfect, are not without their
" use ; as they encourage the execution of
" laborious experiments, or the investigation
" of ingenious deductions to confirm or refute
" them. And since natural objects are allied
" to each other by many affinities, every
" kind of theoretic distribution of them adds
" to our knowledge by developing some of
" their analogies."

Dr. Darwin proved himself more ready to
admit those new and grand views in chemistry
(a branch of science which always greatly
interested him) which were developed
towards the close of the last century, than
some professed chemists. James Keir, a
distinguished chemist of the day, writing to
him in March 1790, says * : " You are such

* ' Sketch of the Life of James Keir, F.R.S.' p. 111.

" an infidel in religion that you cannot believe
" in transubstantiation, yet you can believe
" that apples and pears, &c., sugar, oil,
" vinegar, are nothing but water and char-
" coal, and that it is a great improvement in
" language to call all these things by one
" word —oxyde hydro-carbonneux."

There is a good deal of psychology in the
' Zoonomia,' but I fear that his speculations on
this subject cannot be held to have much
value. Nevertheless, G. H. Lewes says of
him *: "Although even more neglected than
" Hartley by the present generation, Darwin,
" once so celebrated, deserves mention here
" as one of the psychologists who aimed at
" establishing the physiological basis of mental
" phenomena." And again: "Had Darwin
" left us only the passage just cited † we
" should have credited him with a profounder
" insight into psychology than any of his
" contemporaries and the majority of his suc-
" cessors exhibit; and although the perusal of
" ' Zoonomia ' must convince everyone that
" Darwin's system is built up of absurd hypo-

---

* 'History of Philosophy,' 3rd ed. 1867, vol. ii. p. 356.
† 'Zoonomia,' vol. i. p. 27.

" theses, Darwin deserves a place in history
" for that one admirable conception of psy-
" chology as subordinate to the laws of life."
The illustrious Johannes Müller quotes with
approbation, though with correction, his ' law
of associated movements.' *

The ' Zoonomia' is largely devoted to
medicine, and my father thought that it had
much influenced medical practice in England ;
he was of course a partial, yet naturally a more
observant judge than others on this point. The
book when published was extensively read by
the medical men of the day, and the author
was highly esteemed by them as a practitioner.
The following curious story, written down by
his daughter, Violetta, in her old age, shows
his repute as a physician. A gentleman in
the last stage of consumption came to Dr.
Darwin at Derby, and expressed himself to
this effect : " I am come from London to con-
" sult you, as the greatest physician in the
" world, to hear from you if there is any hope
" in my case ; I know that my life hangs upon
" a thread, but while there is life there may

* Müller's ' Elements of Physiology,' translated by Baly, 1842,
p. 943.

" be hope. It is of the utmost importance
" for me to settle my worldly affairs immedi-
" ately; therefore I trust that you will
" not deceive me, but tell me without hesi-
" tation your candid opinion."   Dr. Darwin
felt his pulse, and minutely examined him,
and said he was sorry to say there was no
hope.   After a pause of a few minutes the
gentleman said: " How long can I live?"
The answer was : " Perhaps a fortnight." The
gentleman seized Dr. Darwin's hand and said:
" Thank you, doctor, I thank you; my mind
" is satisfied ; I now know there is no hope for
" me."   Dr. Darwin then said : " But as you
" come from London, why did you not consult
" Dr. Warren, so celebrated a physician?"
" Alas! doctor, I am Dr. Warren."   He died
in a week or two afterwards.

I remember only two points, with respect
to which my father thought that medical
practice in this country had been influenced
by the ' Zoonomia.' *   In this work it is said:
" There is a golden rule by which the neces-

* ' Zoonomia,' 1794, vol. i. p. 99.   I was led to search for
this passage by its having been given by Dr. Dowson in his
' Erasmus Darwin: Philosopher, Poet, and Physician,' 1861,
p. 46.

" sary and useful quantity of stimulus in
" fevers with debility may be ascertained.
" When wine or beer is exhibited, either
" alone or diluted with water, if the pulse
" becomes slower the stimulus is of a proper
" quantity, and should be repeated every two
" or three hours, or when the pulse again
" becomes quicker." The value of this
" golden rule" will be appreciated when it
is remembered that the high importance of
stimulants in fever has only rather recently
been recognised and acted on. His views
on fever certainly attracted attention at the
time ; * but the use of stimulants in such
cases has fluctuated much, and the history
of the subject is an obscure one, as I infer
from a letter which Sir Robert Christison
has had the kindness to send me.

The second point mentioned by my father,
was the treatment of the insane. After saying † that no lunatic should be restrained
unless he be dangerous, Dr. Darwin urges
that in some cases " confinement retards

* See, for instance, Dr. Baeta's work, 'Comparative View of
the Theories and Practice of Drs. Cullen, Brown, and Darwin' :
published in 1800.

† 'Zoonomia,' vol. ii. 1796, p. 352.

" rather than promotes their cure, which is
" forwarded by change of ideas, &c." He
then remarks that mistaken ideas do not by
themselves justify confinement, and adds:
" If everyone who possesses mistaken ideas,
" or who puts false estimates on things, was
" liable to confinement, I know not who of
" my readers might not tremble at the sight
" of a madhouse."

In connection with this subject, the following quotation from Dr. Maudsley is interesting : * " Here I may fitly take occasion to
" adduce certain observations with regard
" to the striking manner in which diseased
" action of one nervous centre is sometimes
" transferred suddenly to another, a fact
" which, though it has lately attracted new
" attention, was long since noticed and com-
" mented on by Dr. Darwin : ' In some con-
" ' vulsive diseases,' he writes, ' a delirium
" ' or insanity supervenes, and the convul-
" ' sions cease ; and conversely, the convul-
" ' sions shall supervene, and the delirium
" ' cease. Of this I have been a witness

---

* ' Patholgy of Mind,' 1879, p. 229.

" ' many times a day, in the paroxysms of
" ' violent epileptics, which evinces that one
" ' kind of delirium is a convulsion of the
" ' organs of sense, and that our ideas are
" ' the motions of those organs.' "

Dr. Lauder Brunton has mentioned to me
another instance in which Dr. Darwin apparently anticipated a modern discovery.

In an article in the ' British Medical Journal '
(1873, p. 735) on " catching cold," Dr. Brunton gives an account of Rosenthal's experiments, showing that when an animal is
exposed to a rather high temperature, " the
" cutaneous vessels become paralysed by the
" heat, and remain dilated even after cold
" has been applied. The blood is thus ex-
" posed over a large surface, and becomes
" rapidly cooled." For instance, the blood
of an animal thus treated fell from between
107·6° and 111° to 96·8°, and remained at
this lower temperature for several days. A
passage in the Zoonomia * seems to show
that Dr. Darwin was acquainted with the
above important fact, discovered by Rosenthal some hundred years later.

* ' Zoonomia,' vol. ii. 1796, p. 570.

Dr. Darwin fully recognised the truth and importance of the principle of inheritance in disease. He remarks : * " As many families " become gradually extinct by hereditary dis- " eases, as by scrofula, consumption, epilepsy, " mania, it is often hazardous to marry an " heiress, as she is not unfrequently the last of " a diseased family." His grandson, Francis Galton, so well known for his works on the subject of inheritance, would fully appreciate this remark. On the other hand, when a tendency to disease is confined to one parent, the children often escape. " I now know," as he writes to my father, January 5th, 1792, " many families who had insanity on *one* side, " and the children, now old people, have had " no symptom of it. If it were otherwise, " there would not be a family in the king- " dom without epileptic, gouty, or insane " people in it."

In ' The Temple of Nature ' (Notes, p. 11), there is a curious instance of his prophetic sagacity with respect to " microscopic ani- " mals." A few years since a utilitarian philo-

* 'The Temple of Nature,' 1803, notes, p. 45; published after his death.

sopher might have sneered at men spending their lives in the examination of organisms far too minute to be seen by the naked eye; and it would have been difficult to have given a satisfactory answer, except on general principles, to such a man. But we now know from the researches of various naturalists how all-important a part these organisms play in putrefaction, fermentation, infectious diseases, &c.; and as a consequence of such researches, the world owes a deep debt of gratitude to Mr. Lister for his anti-septic treatment of wounds. Therefore the following sentence of my grandfather, considering how little was then known on the subject, appears to me remarkable. He says: " I hope that micro- " scopic researches may again excite the at- " tention of philosophers, as unforeseen ad- " vantages may probably be derived from " them like the discovery of a new world."

The 'Phytologia,' was published in 1800. It begins with a discussion on the nature of leaf-buds and flower-buds; and the view, now universally adopted, that a plant consists of "a " system of individuals," and not merely of a multiplication of similar organs, originated with Darwin, as I infer from Johannes Müller's ' Elements of Physiology.'*

Considering how recently the manner in which plants modify and absorb the nutriment stored up in their roots, tubers, cotyledons, &c. has been understood, the following sentence ('Phytologia,' p. 77) deserves notice : " The " digestive powers of the young vegetable, " with the chemical agents of heat and mois- " ture, convert the starch or mucilage of the " root or seed into sugar for its own nourish- " ment ; . . . and thus it appears pro- " bable that sugar is the principal nourish- " ment of both animal and vegetable beings."

The work treats largely of agriculture and horticulture, and a section is devoted to phos- phorus, which, as he believes (p. 207), exists universally in vegetables, a question "which " has not yet been sufficiently attended to." He then refers to the use of bones as a manure, but erred in supposing that shells and some other substances which are luminous in the dark, abounded with phosphorus. Sir J. Sinclair, President of the Board of Agriculture, and

* ' Elements of Physiology,' translated by Baly, 1842, p. 1421.

therefore a most capable judge, says that, though the fertilising properties of bone-dust had been previously noticed by Hunter, yet " they were first theoretically explained and " brought forward with authority by Dr. Dar- " win." He then remarks, and of the truth of his remark there can be little doubt, " per- " haps no (other) modern discovery has contri- " buted so powerfully to improve the fertility " and to increase the produce of the soil."*

The following sentences are interesting as forecasting the progress of modern thought. In a discussion on " The Happiness of Or- ganic Life" (p. 556), after remarking that animals devour vegetables, he says : " The " stronger locomotive animals devour the " weaker ones without mercy. Such is the " condition of organic nature ! whose first law " might be expressed in the words, ' eat or be " eaten,' and which would seem to be one " great slaughter-house, one universal scene " of rapacity and injustice." He proceeds: " Where shall we find a benevolent idea to " console us amid so much apparent misery ?"

* I am indebted to Dr. Dowson's ' Life of Erasmus Darwin,' for the reference to the ' Life and Works of Sir J. Sinclair.'

He then argues : " Beasts of prey more easily
" catch and conquer the aged and infirm, and
" the young ones are defended by their
" parents. . . . . . By this contrivance more
" pleasurable sensation exists in the world
" . . . . old organisations are transmigrated
" into young ones . . . . death cannot so
" properly be called positive evil as the ter-
" mination of good." There is much more of
the same kind, and hardly more relevant.
He then makes a great leap in his argument,
and concludes that all the strata of the world
" are monuments of the past felicity of organ-
" ised nature ! and consequently of the bene-
" volence of the Deity ! "

It is a curious proof of the degree to which
English botanists had been blinded by the
splendour of the fame of Linnæus, that Dr.
Darwin apparently had never heard of Jussieu,
for he writes (p. 564) : " If the system of the
" great Linnæus can ever be intrinsically im-
" proved, I am persuaded that the plan here
" proposed of using the situations, propor-
" tions, or forms, with or without the number
" of the sexual organs, as criterions of the
" orders and classes, must lay the foundation ;

" but that it must require a great architect
" to erect the superstructure." He therefore
did not know that a noble superstructure had
already been raised.

There remains only one other book to
be noticed : ' A Plan. for the Conduct of
Female Education in Boarding Schools,' published in 1797. This is a short treatise which
seems never to have received much attention
in England, though it was translated into
German. It is strongly characterised by plain
common sense, with little theorising, and is
throughout benevolent. He insists that
punishment should be avoided as much as
possible, and that reproof should be given
with kindness. Emulation, though useful, is
dangerous, from being liable to degenerate
into envy. "If once you can communicate
" to children a love of credit and an appre-
" hension of shame, you have instilled into
" them a principle, which will constantly act
" and incline them to do right, though it is
" not the true source whence our actions
" ought to spring, which should be from our
" duty to others and ourselves." He urges
that sympathy with the pains and pleasures

of others is the foundation of all our social
virtues; and that this can best be inculcated
by example and the expression of our own
sympathy. "Compassion, or sympathy with
" the pains of others, ought also to extend to
" the brute creation . . . to destroy even
" insects wantonly, shows an unreflecting
" mind, or a depraved heart."

He considers it of great importance to girls
that they should learn to judge of character,
as they will some day have to choose a husband; and he believes that reading proper
novels teaches them something of life and
mankind, and helps them to avoid mistakes
in judging of character. He also remarks
more than once, that children express various
emotions in their countenances much more
plainly than older persons; and he is convinced that one great advantage which a child
derives from going to school is in uncon-
sciously acquiring a knowledge of physi-
ognomy through mixing with other children.
This knowledge, " by giving a promptitude
" of understanding the present approbation
" or dislike, and the good or bad designs of
" those whom we converse with, becomes of

" hourly use in almost any department of
" life."

He was much in advance of his age in his
ideas as to sanitary arrangements—such as
supplying towns with pure water, having
holes made into crowded sitting and bed-
rooms for the constant admission of fresh air,
and not allowing chimneys to be closed during
summer, and as to diet and exercise. He
speaks of " skating on the ice in winter,
" swimming in summer, funambulation or
" dancing on the straight rope," as " not
" allowed to ladies by the fashion of this age
" and country." It is a pity he does not tell
us when and where it was the fashion of
young ladies to funambulate! With respect
to swimming, he disregarded fashion, and had
his own daughters as well as his sons taught
to swim at a very early age, so that they
became, it is said, expert swimmers as early
as four years old. In the 'Phytologia' he
shows himself still more clearly a great
sanitary reformer. He insists that the sewage
from towns, which is now left buried or
carried into the rivers, should be removed for
the purpose of agriculture; "and thus the

" purity and healthiness of the towns may
" contribute to the thriftiness and wealth of
" 'the surrounding country." " There should
" be no burial places in churches or in church-
" yards, where the monuments of departed
" sinners shoulder God's altar, . . . but
" proper burial grounds should be consecrated
" out of towns." Nearly a century has
elapsed since this good advice was given,
and it has as yet only partially been carried
out.

One of the subjects which interested Dr.
Darwin most throughout his whole life, and
which appears little in his published works, was
mechanical invention. This is shown in his
letters to Josiah Wedgwood, Edgeworth, and
others, and in a huge common-place book full
of sketches and suggestions about machines.
He seems, however, rarely to have completed
anything, with the exception of a horizontal
windmill for grinding flints, which he designed for Wedgwood, and which answered
its purpose. There are schemes and sketches
for an improved lamp, like our present
moderators ; candlesticks with telescope stands

so as to be raised at pleasure to any required
height; a manifold writer; a knitting loom for
stockings; a weighing machine; a surveying
machine; a flying bird, with an ingenious
escapement for the movement of the wings,
and he suggests gunpowder or compressed air
as the motive power. He also gives a plan of
a canal lock, on the principle of the boat being
floated into a large box, the door of which is
then closed, and the box afterwards raised or
lowered. This principle has since been acted
on under certain circumstances, but by an
improved method. A rotatory pump was
also one of his schemes, and this, under a
modified form, is extensively used for blowing
air into smelting cupolas, and for pumping
water in certain cases. He saw clearly, as he
explains in 1756 in a letter to Reimarus, that
it would be a great advantage if the spokes of
carriage wheels acted as springs; and Sir J.
Whitworth has recently had a carriage constructed with such wheels, which is remarkably smooth.

Another invention was a small carriage of
peculiar construction, intended to give the
best effect to the power of the horse, combined

with the greatest ease in turning. " It was
" a platform," says Miss Seward, " with a
" seat fixed upon a very high pair of wheels,
" and supported in the front upon the back
" of the horse, by means of a kind of pro-
" boscis, which, forming an arch, reached over
" the hind quarters of the horse ; and passed
" through a ring, placed on an upright piece
" of iron, which worked in a socket, fixed in
" the saddle."*   But however correct this
carriage may have been in principle,. Darwin
had the misfortune, in the year 1768, to be
upset in it, when he broke his knee-cap and
ever afterwards limped a little.

A speaking machine was a favourite idea,
and for this end he invented a phonetic alphabet. His machine, or ". head, pronounced the
" *p*, *b*, *m*, and the vowel *a*, with so great
" nicety as to deceive all who heard it unseen,
" when it pronounced the words *mama*, *papa*,
" *map*, and *pam*; and it had a most plain-

---

* Dr. Krause informs me " that the Moravian engineer, Theodo
" Tomatschek, has lately constructed a very similar carriage,
" which I saw at the Vienna International Exhibition ; and the
" Americans have also reduced the Darwinian idea to practice,
" and given the new vehicle the paradoxical name ' Equibus.'"

" tive tone, when the lips were gradually
" closed."* Edgeworth also bears witness to
the capacity of this speaking head. Matthew
Boulton entered into the following agreement,
which, from the witnesses to it, was evidently
made at one of the meetings of the famous
Lunar Club ; but whether in joke or earnest,
it is difficult to conjecture :

I promise to pay to Dr. Darwin of Lichfield
one thousand pounds upon his delivering to me
(within 2 years from date hereof) an Instrument
called an organ that is capable of pronouncing the
Lord's Prayer, the Creed, and Ten Commandments
in the Vulgar Tongue, and his ceding to me, and me
only, the property of the said invention with all the
advantages thereunto appertaining.

M. Boulton
Soho Sep. 3rd 1877
Witness, James Keir
Witness, W. Small

In the last century a speaking tube was an
unknown invention in country districts, and

* 'Temple of Nature,' notes, p. 120; p. 107 on the phonetic
alphabet. See also 'Memoirs of Edgeworth,' vol. ii. p. 178.

Dr. Darwin had one for his study, which opened near the back of the kitchen fireplace. A countryman had brought a letter and sat waiting for an answer by this fire, which had become very low, when suddenly he heard a sepulchral voice, saying, as if from the depths of the expiring fire, " I want " some coals." The man instantly fled from the house, for my grandfather had the reputation amongst the country folk of being a sort of magician.

At a time (1783) when very few artesian wells had been made in this country, Dr. Darwin made one, though on a small scale; and in the garden-wall to his house in Full Street, Derby, there still exists an iron plate with the following inscription :

TERREBELLO EDUXIT AQUAM
ANNO MDCCLXXXIII.
ERASMUS DARWIN.
LABITUR ET LABETUR.

This case would not have been worth mentioning had he not shown in his paper,* in

* 'Philosophical Transact.' 1785, part i. p. 1.

which this well is described, that he recognised the true principle of artesian wells.
He remarks that "some of the more interior
" strata of the earth are exposed naked on
" the tops of mountains ; and that in general,
" those strata which lie uppermost, or nearest
" to the summit of the mountain, are the
" lowest in the contiguous plains." He then
adds that the waters "sliding between two
" of the strata above described, descend till
" they find or make for themselves an outlet,
" and will in consequence rise to a level
" with the parts of the mountain where they
" originated."

In Oct. 1771 he wrote several letters to
Wedgwood about a scheme of making, with
his own capital, a canal of very small dimen-
sions from the Grand Trunk to Lichfield, for
boats drawing only a foot of water, to be
dragged by a man, and carrying only four or
five tons burthen. Such a canal would have
borne the same relation to ordinary canals, as
some very narrow railways, which have been
found to answer well in Wales, bear to ordinary
railways. He seems to have been greatly

interested in this project, which, however, never came to anything.

The weather, and the course of the winds throughout the world, was another subject on which he was continually searching for information and speculating. I have heard my father say, that in order to notice every change of the wind he connected a wind-vane on the top of his house with a dial on the ceiling of his study.

There remains only to be said that Erasmus Darwin died at Breadsall Priory, near Derby, on Sunday morning, April 18th, 1802, in his seventy-first year. A week previously he had been ill for a few days, but had recovered. On the 17th, whilst walking in his garden with a lady, he told her that he did not expect to live long. At night he was as cheerful as usual. On the following morning, the 18th, he rose at six o'clock and wrote a long letter to Mr. Edgeworth * which he did not live to finish, and which contains the following description of the Priory, where he had

* R. L. Edgeworth's 'Memoirs,' 2nd ed., vol. ii. p. 242.

BREADSALL PRIORY, WHERE ERASMUS DARWIN DIED.

been living for about two years : " We have
" a pleasant house, a good garden, ponds full
" of fish, and a pleasing valley somewhat like
" Shenstone's—deep, umbrageous, and with a
" talkative stream running down it. Our
" house is near the top of the valley, well
" screened by hills from the east and north,
" and open to the south, where, at four miles
" distance, we see Derby tower." At about
seven o'clock he was seized with a violent
shivering fit, and went into the kitchen to
warm himself; he returned to his study, lay
on the sofa, became faint and cold, and was
moved into an arm-chair, where without pain
or emotion of any kind he expired a little
before nine o'clock.

A few years before he had written to Edgeworth: "When I think of dying, it is always
without pain or fear ;" but he had often expressed a strong hope that his end might be
painless, and so it proved. His medical attend-
ants differed about the cause of his death, but
my father did not doubt that it was an affection of the heart. Many years afterwards his
widow showed me the sofa and chair, still
preserved in the same place, where he had lain

and expired. He was buried in Breadsall Church.

ERASMUS DARWIN, M.D., F.R.S.
Born at Elston, near Newark, 12th Dec., 1731.
Died at the Priory, near Derby, 10th April, 1802.
Of the rare union of Talents
which so eminently distinguished him
as a Physician, a Poet and Philosopher
His writings remain
a public and unfading testimony.
His Widow
has erected this monument
in memory of
the zealous benevolence of his disposition,
the active humanity of his conduct,
and the many private virtues
which adorned his character.

THE SCIENTIFIC WORKS

OF

ERASMUS DARWIN.

By ERNST KRAUSE.

TRANSLATED FROM THE GERMAN BY W. S. DALLAS, F.L.S.

K

On the second page of the later editions of Darwin's 'Origin of Species' * we find the following brief observation :—" It is curious how largely my grandfather, Dr. Erasmus Darwin, anticipated the views and erroneous grounds of opinion of Lamarck in his 'Zoonomia' (vol. i. pp. 500–510), published in 1794." Being quite aware of the reticence and modesty with which the author expresses himself, especially in speaking *pro domo*, I thought immediately that here we ought to read between the lines, and that this ancestor of his must certainly deserve considerable credit in connection with the history of the Darwinian theory. As no light was to be obtained upon this subject from German literature I procured the works of Erasmus Darwin, and have found singular pleasure in their study.

I was speedily convinced that this man,

* Sixth edition, p. xiv. note.

equally eminent as philanthropist, physician,
naturalist, philosopher, and poet, is far less
known and valued by posterity than he
deserves, in comparison with other persons
who occupy a similar rank. It is true that
what is perhaps the most important of his
many-sided endowments, namely his broad
view of the philosophy of nature, was not
intelligible to his contemporaries; it is only
now, after the lapse of a hundred years, that
by the labours of one of his descendants we
are in a position to estimate at its true value
the wonderful perceptivity, amounting almost
to divination, that he displayed in the domain
of biology. For in him we find the same in-
defatigable spirit of research, and almost the
same biological tendency, as in his grandson;
and we might, not without justice, assert that
the latter has succeeded to an intellectual
inheritance, and carried out a programme
sketched forth and left behind by his grandfather.

Almost every single work of the younger
Darwin may be paralleled by at least a
chapter in the works of his ancestor; the
mystery of heredity, adaptation, the pro-

tective arrangements of animals and plants,
sexual selection, insectivorous plants, and
the analysis of the emotions and sociological
impulses; nay, even the studies on infants are
to be found already discussed in the writings
of the elder Darwin. But at the same time
we remark a material difference in their inter-
pretation of nature. The elder Darwin was a
Lamarckian, or, more properly, Jean Lamarck
was a Darwinian of the older school, for
he has only carried out further the ideas
of Erasmus Darwin, although with great
acumen; and it is to Darwin therefore that
the credit is due of having first established a
complete system of the theory of evolution.
The evidence of this I shall adduce hereafter.

The unusual circumstance that a grandfather should be the intellectual precursor of
his grandson in questions which now-a-days
more than any others move the minds of men,
must of itself suffice to excite the liveliest
interest. But at the same time it must be
pointed out that in this fact we have not
the smallest ground for depreciating the
labours of the man who has shed a new
lustre upon the name of his grandfather. It

is one thing to establish hypotheses and
theories out of the fulness of one's fancy,
even when supported by a very considerable
knowledge of nature, and another to demonstrate them by an enormous number of facts,
and carry them to such a degree of probability
as to satisfy those most capable of judging.
Dr. Erasmus Darwin could *not* satisfy his
contemporaries with his physio-philosophical
ideas; he was a century ahead of them, and
was in consequence obliged to put up with
seeing people shrug their shoulders when
they spoke of his wild and eccentric fancies,
and the expression " Darwinising " (as employed for example by the poet Coleridge
when writing on Stillingfleet) was accepted
in England nearly as the antithesis of sober
biological investigation.*

The many-sidedness of his endowments also
injured his fame in another direction. The
physicians reproached him with being a
philosopher; and the philosophers thought
themselves justified in complaining that he
was of far too poetical and fanciful a constitution; the poets and *literati* on the other

* See 'Athenæum,' March, 1875, p. 423.

hand objected to his position as a physician and his scientific tendencies; and thus partiality and prejudice prevented his judges from a full and complete recognition of the value of the man. His life and labours have frequently been described, but always by either *littérateurs* or medical men, and hence the picture produced has always had a partizan colouring.

Nevertheless it is gratifying to find that each of his biographers has expressed the highest appreciation of precisely that side of the doctor's activity of which he was most capable of judging. The *literati* formerly extolled his poetical merits. Eighteen years ago an English physician praised his medical contributions; and it has remained for the present writer to add to these the hitherto neglected tribute of recognition* which is due to him on the part of natural history and physio-philosophy.

It is characteristic of this distinguished man that he never exhibited those fluctuating opinions with respect to the evolution of organic beings which are evident in the works of Linnæus and Buffon.

* See ' Kosmos,' February, 1879, p. 393.

When Göthe, in the year 1786, penetrated
by the thought that a common organization
must bind together the higher animals, demonstrated the existence of the intermaxillary
bone in man, the supposed absence of which
had been regarded as a character clearly
separating man from animals, no anatomist
would agree with him; his idea of vegetable
metamorphosis, which he brought forth about
the same time, was strenuously opposed by the
botanists; and his discovery in 1790 of the
vertebral nature of the skull has only met
with justice in our own days. Exactly similar
was the fate of Dr. Darwin, who, as we shall
show, was far in advance of his age. Exceedingly successful in grasping and combining
separated things, Göthe absolutely detested
the analytical activity of the exact investigator,
although he availed himself of it, and indeed
exercised it himself in procuring the materials
for his new conception of the world. Dr.
Darwin had no such aversion to the analytical
activity of the philosophers and specialists,
and hence he carried his construction further
than any of his predecessors and contemporaries. The similarity of the conceptions

of the universe of the two poets is in many respects as great as their need to give utterance to them in verse; but this agreement may be easily explained if we consider that both of them started from the investigations of the same precursors, Buffon and Linnæus.

The first great work of Darwin, the didactic poem ' The Botanic Garden,' is divided into two parts, which are not very closely connected; for this reason I shall hereafter cite the second part, ' The Loves of the Plants,' which appeared before the first, under the above special title.* The first part, ' The Economy of Vegetation,' certainly answers to both the principal and special titles only in its last canto, the first three cantos describing the action of the forces of nature in general, and specially the formation of the world. Various critics have expressed the opinion that Dr. Darwin's didactic poem was an imitation of one which appeared anonymously in London in 1735 under the title of ' Universal Beauty,' the author of which afterwards turned out to

---

* The following citations refer throughout to the second editions, both of the first part (' The Economy of Vegetation,' London, Johnson, 1791) and of the second (' The Loves of the Plants,' London, Nichols, 1790).

be the poet Henry Brooke. Others have represented Sir Richard Blackmore's poem, ' The
Creation,' which appeared in 1712, as the
model.* Neither statement has the slightest
foundation. Henry Brooke's ' Universal
Beauty' is a " Physico-theology " in verse,
which, although decidedly more sonorous and
poetical than the offspring of the similarly
employed muse of his German namesake
(Heinrich Brookes), is merely devoted to a
representation of the glories of creation of
the same character as the physico-theologies of
that period. Blackmore's 'Creation,' which,
from its being divided into seven books, people
have been led to regard as belonging to the
Diluvianistic literature, treats of the process of
creation only by the way; and is essentially a
purely polemico-rhetorical philippic against the
atheists, from Democritus and Epicurus down
to Descartes and Spinoza, in which we find so
little sound judgment and insight that the
author can by no means make up his mind

---

* The suggestion that Dr. Darwin may have made use of
Brooke's ' Universal Beauty ' as his pattern, seems to have first
appeared in a critical article in the ' Edinburgh Review ' (April,
1803, 4th ed. p. 491), but has since passed, as a demonstrated fact,
into later biographical works, e.g., the ' Biographie Universelle.'

whether he shall decide in favour of Aristotle and Ptolemy, or of Copernicus, Kepler, and Newton. The German critics who regard Blackmore's poem as the model of Darwin's 'Botanic Garden' must certainly have neglected to read at least one of these didactic poems. Blackmore's work might much rather be regarded as the pattern for Polignac's 'Anti Lucretium,' although it is far exceeded by the latter in dialectic acuteness.

In the introduction and apology to the 'Botanic Garden' the author says: "The "general design of the following sheets is to "inlist Imagination under the banner of "Science; and to lead her votaries from the "looser analogies which dress out the imagery "of poetry, to the stricter ones, which form "the ratiocination of philosophy. . . . It "may be proper here to apologize for many "of the subsequent conjectures on some "articles of natural philosophy, as not being "supported by accurate investigation or con- "clusive experiments. Extravagant theories, "however, in those parts of philosophy, where "our knowledge is yet imperfect, are not "without their use; as they encourage the

" execution of laborious experiments, or the
" investigation of ingenious deductions, to
" confirm or refute them."

The plan of the poem was to a certain
extent prescribed by those initial verses of
Miss Seward's, which the author placed at the
commencement of his work, either out of
gallantry or in acknowledgment of their
having given the first inducement to the
production of the poem. Starting from the
fundamental idea that the mythology of the
ancients glorified the forces and government
of nature in the persons of their deities,
he has introduced the personified forces of
nature which prevail in fire, air, water, and
earth; and then represents the goddess as
addressing herself to the different groups of
elementary spirits, in a figurative discourse,
permeated throughout with mythological elements, and describing the part taken by each
in the formation and life of the world. Thus
the first canto is addressed to the " nymphs of
primeval fire," and he accordingly describes
the production of the universe from this
source, at the same time bringing together
many of the ordinary phenomenal forms of

fire, heat, and light. Matters which can only
be slightly touched upon in the verses are
further elaborated, partly in short footnotes
and partly in more detailed memoirs (additional notes) which are relegated to the end
of the volume. It is to these notes that our
attention must principally be directed.

We are especially interested in a note to
verse 101 of the first canto, in which the
author unfolds the idea and the first scheme
of the theory of evolution. He says: "From
" having observed the gradual evolution of the
" young animal or plant from its egg or seed;
" and afterwards its successive advances to its
" more perfect state, or maturity; philosophers
" of all ages seem to have imagined that the
" great world itself had likewise its infancy and
" its gradual progress to maturity; this seems
" to have given origin to the very antient
" and sublime allegory of Eros, or Divine love,
" producing the world from the egg of Night,
" as it floated in chaos." To the second
particularly important part of this note we
shall have to refer hereafter.

For the student of the history of civilization
who looks back from a Darwinistic standpoint,

a fancy worked out in this canto as to the
discovery and subjugation of fire, which
Darwin denominates " the first art," will be
particularly interesting.

> " Nymphs ! your soft smiles uncultur'd man subdued,
> And charm'd the Savage from his native wood ;
> You, while amaz'd his hurrying Hords retire
> From the fell havoc of devouring Fire,
> Taught, the first Art ! with piny rods to raise
> By quick attrition the domestic blaze,
> Fan with soft breath, with kindling leaves provide,
> And list the dread Destroyer on his side.
> So, with bright wreath of serpent-tresses crown'd,
> Severe in beauty, young Medusa frown'd ;
> Erewhile subdued, round Wisdom's Ægis roll'd,
> Hiss'd the dread snakes, and flam'd in burnish'd gold ;
> Flash'd on her brandish'd arm the immortal shield,
> And Terror lighten'd o'er the dazzled field." *

We then have the well-known verses on
the power of steam, vv. 289–296.

> " Soon shall thy arm, Unconquer'd Steam, afar
> Drag the slow barge, or drive the rapid car ;
> Or on wide-waving wings expanded bear
> The flying-chariot through the fields of air.
> —— Fair crews, triumphant, leaning from above,
> Shall wave their flutt'ring kerchiefs as they move ;
> Or warrior-bands alarm the gaping crowd,
> And armies shrink beneath the shadowy cloud.

* ' Economy of Vegetation,' canto i. vv. 209–222.

So mighty Hercules o'er many a clime
Waved his vast mace in Virtue's cause sublime,
Unmeasured strength with early art combined,
Awed, served, protected, and amazed mankind."

The second canto is addressed to the gnomes or earth-spirits, and describes the gradual development of the earth, which, with the other planets, the author believes to have been cast forth from a volcano in the sun. By stronger friction or adhesion to one wall of this volcano the earth received its axial revolution and spheroidal form; by refrigeration a nucleus was formed, upon which the waters were precipitated as a primeval ocean free from salt, while the lighter gases formed the atmosphere. " It is probable," he adds, " that all " the calcareous earth in the world . . . . . " was formed originally by animal and vege- " table bodies from the mass of water."* By the lixiviation of the rocks the seas became salt. Finally the formation of the vegetable

* It was a favourite notion of Dr. Darwin's that all the lime of the earth originated from living creatures, corals, shells, and other animals, and therefore must have taken part in the pleasures and pains of life. The limestone mountains of England appeared to him as " mighty monuments of past delight." It was probably in consequence of this idea, and in allusion to his family arms, consisting of three scallop shells, that he altered his motto to " E conchis omnia."

world is indicated, to which may be added here from the second part (pp. 36 and 44) that Darwin regarded lichens as the oldest terrestrial plants, and, like Häckel in more recent times, he referred the fungi to a kingdom which, like "a narrow isthmus," united plants and animals.

In the third canto, addressed to the water-nymphs, the circulation and action of water upon the earth is described. The formation of clouds, the sea and its life, springs, rivers, geysers, glaciers, coral structures, &c. In this connection the fossil marine animals also come under discussion; and after mentioning the singular circumstance that most fossil marine animals as, for example, the ammonites, are no longer found living, whilst the living animals do not occur in the fossil state, the author raises the questions, " Were all the " ammoniæ destroyed when the continents " were raised ? Or do some genera of animals " perish by the increasing power of their " enemies ? Or do they still reside at in- " accessible depths in the sea ? Or do some " animals change their forms gradually and " become new genera ?"*

* 'The Economy of Vegetation,' p. 120.

The question of the transformation of species and their development into higher forms was a favourite one with the elder Darwin, and one to which he has given expression in all his works, at least in one place, and usually in very similar terms. Already, in the eighth page of the poem now under consideration, he brings it forward, and after having spoken of the stratified formation of the earth in a note, the commencement of which has already been given, he says: " There are likewise " some apparently useless or incomplete " appendages to plants and animals which " seem to shew they have gradually under- " gone changes from their original state ; such " as the stamens without anthers, and styles " without stigmas of several plants, as men- " tioned in the note on Curcuma, vol. ii. of " this work. Such as the halteres, or rudi- " ments of wings of some two-winged insects, " and the paps of male animals ; thus swine " have four toes, but two of them are im- " perfectly formed, and not long enough for " use." We here break off in order to append the above-mentioned note on the Turmeric

L

plant, which gives the theory of rudimentary
organs still more completely. "There is a
" curious circumstance," he says, " belonging
" to the class of insects which have two wings,
" or diptera, analogous to the rudiments of
" stamens above described; viz. two little
" knobs are found placed each on a stalk or
" peduncle, generally under a little arched
" scale; which appear to be rudiments of
" hinder wings; and are called by Linneus
" halteres, or poisers, a term of his introduc-
" tion. Other animals have marks of having
" in a long process of time undergone changes
" in some parts of their bodies, which may
" have been effected to accommodate them to
" new ways of procuring their food. The
" existence of teats on the breasts of male
" animals, and which are generally replete
" with a thin kind of milk at their nativity, is
" a wonderful instance of this kind. Perhaps
" all the productions of nature are in their
" progress to greater perfection?—an idea
" countenanced by the modern discoveries
" and deductions concerning the progressive
" formation of the solid parts of the ter-

" raqueous globe, and consonant to the dignity
" of the Creator of all things."*

Buffon before him had regarded the rudimentary organs somewhat in the same way,
but he had by no means perceived with equal
clearness their part as evidence in favour
of the theory of descent. The pig, says
Buffon, rather mysteriously, " does not appear
" to have been formed upon an original,
" special and perfect plan, since it is a com-
" pound of other animals ; it has evidently
" useless parts, or rather parts of which it
" cannot make any use,—toes, all the bones
" of which are perfectly formed, and which
" nevertheless are of no service to it. Nature
" is consequently far from subjecting herself
" to final causes in the formation of her
" creatures. Why should she not sometimes
" add superfluous parts, when she so often
" seems to omit essential ones ? . . . . Why
" do we regard it as necessary that in each
" individual every part should be useful to
" the others and necessary to the whole ?
" Does it not suffice for their co-existence that
" they do not injure one another, that they

* 'Loves of the Plants,' pp. 7, 8.

" can grow without hindrance, and develope
" without obliterating each other ?   All parts
" which do not sufficiently injure one another
" to cause mutual destruction, all that can
" exist together, exist; and perhaps in the
" majority of living creatures there are fewer
" related, useful or necessary, than indifferent,
" useless or superfluous parts. But we, always
" wishing to refer everything to a certain
" purpose, when parts have no apparent use,
" invent for them hidden purposes and
" imagine unfounded relations which do not
" exist in the nature of things, and only serve
" to obscure matters.   We fail to see that
" thus we deprive philosophy of its true
" character, and misrepresent its object, which
" consists in the knowledge of the 'How' of
" things, the way in which nature acts, and
" that we substitute for this real object a vain
" idea by seeking to divine the 'Why' of
" the facts, or the purpose which she has in
" her activity." *

Buffon had a dim idea that rudimentary
organs and similar irregularities found their
explanation in the consideration of the general

* ' Hist. Nat.' tome v. 1755, pp. 103, 104.

connection of natural objects; he indicated
that doubtful species, irregular structures, and
anomalous existences found their place in the
eternal order of things, as well as all others,
and that they complete the links of the chain;
but he has not expressed his opinion upon
these points with any distinctness, like Dr.
Darwin. The chief force of the above words
is evidently directed against the physico-
theologians.

The last century was a period of the most
industrious and endless search after design.
In opposition to the French philosophy,
with its materialistic tendency, innumerable
hosts of pious writers came forward in England, Holland, and especially in Germany,
and undertook to prove the divine origin
of all things from the study of nature itself,
and indeed, from every straw and sand-
grain. Following on the two best works of
this kind, namely, Swammerdamm's 'Biblia
Naturæ,' and John Ray's 'The Wisdom of
God Manifested in Creation' (1691), there
poured forth upon the people such a flood
of writings upon natural theology that a
book would be required to give a toler-

able review only of the chief of them :—
Nehemiah Grew's 'Cosmologia Sacra' (1711),
and Derham's ' Astro-,' Physico-,' Hydro-' and
' Pyro-theology' were occupied more with
general questions, but in Germany, on this
field favoured by the Leibnitz-Wolfian philosophy, the minutest details were gone into.
A shallow, sickly enthusiasm, which was
called " natural religion," gained the upper
hand ; the whole world appeared only to
exist for the service, pleasure and edification
of man. Lesser's ' Litho-theologie' (1735)
and Rohr's ' Phyto-theologie' (1739) were
followed, going more into detail, by Lesser's
' Insecto-theologie' (1738), and the same
learned pastor's ' Testaceo-theologie,' Zorn's
' Petino-theologie' (1742) and two ' Ichthyo-
theologies' by Malm and Richter (1751 and
1752). Gradually even the individual species
of animals took their turn, e.g., the bees in
Schierach's ' Melitto-theologie' (1767) ; nay,
even such natural phenomena of very doubtful benefit as swarms of locusts and earth-
quakes were rendered harmless in Rathleff's
voluminous ' Acrido-theologie' (1748) and
Pren's ' Sismo-theologie' (1772). That Hein-

sius celebrated " Snow as an admirable creature of God" in his ' Chiono-theologie '
(1735), and Ahlwardt did the same good
service to thunder and lightning in his
' Bronto-theologie' (1745) was only right and
proper.

Buffon could not escape from this tendency
of his time, and in the first volume of his
' Natural History' he devoted a long justifica-
tory chapter to the mountains which Burnet
had charged with being evidences of the Fall
of Man. Feuerlin, however, had preceded
him with a Latin Dissertation on the mountains as divine witnesses (1729) in opposition
to Lucretius and Burnet.*

Against this movement, to which Brooke's
poem already mentioned also pertains, the
elder Darwin opposed himself, not indeed
expressly, but for that very reason the more
efficaciously. He did not inquire how far
this or that property of plants or animals
was directly or indirectly serviceable to man,
but rather whether particular properties

* This enumeration of physico-theological writings is derived
from the elaborate work of G. Zöckler, ' Geschichte der Bezie-
hungen zwischen Theologie und Naturwissenschaft,' Gütersloh,
1877-79.

were not useful to the organisms themselves, and whether it was conceivable that they could have acquired such properties as favoured their well-being by an internal impulse and gradual improvement. For a time he seems to have addressed to every creature that came before him, some such apparently curious questions as these : Why does any creature have this and no other appearance ? Why has this plant poisonous juices ? Why has that one spines ? Why have birds and fishes light-coloured breasts and dark backs ? &c., &c. The last canto of the first part of the ' Botanic Garden,' and the second part generally, are particularly rich in such justly-raised and truly Darwinistic questions. We shall have to recur to this point hereafter, and now, after this digression, return once more to the analysis of the ' Botanic Garden.'

In the fourth canto, addressed to the sylphs, after some descriptions of winds and climates, the author turns to the daughters of the air, the plants, and describes their economy, in the course of which a great number of exceedingly modern remarks are anticipated. In a note to verse 411 (p. 194) the digestion of the reserve material in the seed-lobes during germination is described as a process perfectly analogous to animal digestion, and for some years we have been aware that this comparison is justified even in its details; but above all, in the second part, in which plants are arranged in accordance with the sexual system, and their several relations especially described in separate pictures, that theme of the protection of plants from unbidden guests, which Kerner three years ago made the subject of an interesting book,* is referred to.

Here we learn in the first place that the waxy and resinous secretions of the green parts serve as protections against cold and moisture; and that essential oils, strong odours, and poisons are useful to plants, by protecting them from marauding insects and other animals. The root of the meadow saffron (*Colchicum autumnale*), which does not ripen its seeds until the following spring, would be in danger of destruction in winter

* 'Die Schutzmittel der Blüthen gegen unberufene Gäste. Vienna, 1876.

by animals living in the ground, if it did not contain so acrid a poison.* This example of a poisonous bulb is particularly instructive, because here, in consequence of the seeds ripening only in the next period of vegetation, the existence of the plant in winter would be seriously compromised if the bulb were edible.

The holly (*Ilex aquifolium*) led Dr. Darwin to specially thoughtful considerations in this direction; he speaks of it as follows :† " Many " plants, like many animals, are furnished " with arms for their protection ; these are " either aculei, prickles, as in rose and bar- " berry, which are formed from the outer " bark of the plant; or spinæ, thorns, as in " hawthorn, which are an elongation of the " wood, and hence more difficult to be torn " off than the former ; or stimuli, stings, as in " the nettles, which are armed with a venom- " ous fluid for the annoyance of naked " animals. The shrubs and trees, which have " prickles or thorns, are grateful food to " many animals, as goosberry and gorse ; and

---

* ' The Loves of the Plants,' p. 22, note.
† *Ibid.* pp. 18, 19, note.

" would be quickly devoured, if not thus
" armed; the stings seem a protection against
" some kinds of insects, as well as the naked
" mouths of quadrupeds. Many plants lose
" their thorns by cultivation, as wild animals
" lose their ferocity, and some of them their
" horns. A curious circumstance attends the
" large hollies in Needwood Forest; they are
" armed with thorny leaves about eight feet
" high, and have smooth leaves above; as if
" they were conscious that horses and cattle
" could not reach their upper branches."

On the other hand, that the plants thus
armed furnish animals with an especially
dainty food is proved by the fondness of the
ass for thistles, and of the horse for furze, of
which the author gives an instructive example in a book which will be noticed hereafter. He says: " In the extensive moorlands
" of Staffordshire the horses have learnt to
" stamp upon a gorse-bush with one of their
" fore-feet for a minute together, and when
" the points are broken, they eat it without
" injury; which is an art other horses in
" the fertile parts of the country do not

" possess, and prick their mouths till they
" bleed, if they are induced by hunger or
" caprice to attempt eating gorse." *

This observer of nature was particularly
interested in the means possessed by plants
for preventing the crawling up of wingless
insects into the flowers. He explained in
this way the small water-basins which the
leaves form about the stem of the Fuller's
Teasel, and which have recently led to a
remarkable investigation on the part of one
of his descendants,† as also the larger basins
which surround the flower stalks of the
Bromeliaceæ, as being arrangements destined
partly to the refreshment of the plants, and
partly to serve as a protection for its flowers
and seeds.‡ A similar protective contrivance
occurs most instructively in the viscous rings
of the catchfly, the description of which
may follow here as a sample of the ' Loves
of the Plants,' with the preliminary remark
that the numbers relate to the stamens and

* ' Zoonomia,' vol. i. p. 162, sect. xvi. ii.
† See ' Kosmos,' i. p. 354.
‡ ' The Loves of the Plants,' p. 37.

styles in each of these individual descriptions.

" The fell Silene and her sisters fair,
  Skill'd in destruction, spread the viscous snare.
  The harlot-band *ten* lofty bravoes screen,
  And frowning guard the magic nets, unseen.
  Haste, glittering nations, tenants of the air,
  Oh, steer from hence your viewless course afar!
  If with soft words, sweet blushes, nods, and smiles,
  The *three* dread Syrens lure you to their toils,
  Limed by their art in vain you point your stings,
  In vain the efforts of your whirring wings!—
  Go, seek your gilded mates and infant hives,
  Nor taste the honey purchas'd with your lives!"

In a note upon this passage of his poem
(pp. 15, 16) Darwin remarks : " The viscous
" material which surrounds the stalks under
" the flowers of this plant, and of the Cucu-
" balus Otites, is a curious contrivance to
" prevent various insects from plundering the
" honey, or devouring the seed. In the
" Dionæa Muscipula there is a still more won-
" derful contrivance to prevent the depreda-
" tions of insects ; the leaves are armed with
" long teeth, like the antennæ of insects, and
" lie spread upon the ground round the stem ;
" and are so irritable, that when an insect
" creeps upon them, they fold up, and crush

" or pierce it to death." The same explanation is satisfactory to him for the capture of
insects by the leaves of the Sundew (*Drosera*),
at the same time, that both plants had been
already suspected of using the captured
insects as food. Diderot, it may be remarked
in passing, appears to have been the first to
employ the expression " carnivorous plants ;"
he said of the Venus' fly-trap (*Dionæa*),
" Voilà une plante presque carnivore."*

We must dwell a little longer upon the
investigations of the elder Darwin upon the
protective arrangements of plants, because
they explain to us a remarkable error into
which this acute naturalist fell with respect
to the secretion of honey in flowers. He
believed, especially from the last-mentioned
examples, that plants were generally equipped
so as to keep insects and other lovers of
honey away from the flowers ; and he was
strengthened in this opinion by the circumstance that the source of honey in most
flowers is very much concealed, and often
hidden under complex protective contrivances. He also thought that the resemblance

* Diderot, Œuvres, ed. d'Assézat, tome xi. p. 227.

of the flowers of many orchids to insects could be best explained by a sort of mimicry. His idea, which was very ingenious although fallacious, was that they had acquired the aspect of flowers already occupied by insects in order to be protected from the visits of lovers of honey. Thus the flowers of the Fly-Ophrys resemble a small wall-bee (*Apis ichneumonea*) so closely that at a small distance they appear to be already occupied; and a South-American Cypripedium even resembles the bird-catching spider, in order to frighten away the humming birds, which are so greedy of honey.* Although founded on false examples, the principle of mimicry is here quite correctly expounded, and perhaps for the first time.

The works of Kölreuter † (1761) and Sprengel (1793), which explained the contrivances for the allurement of insects, appear to have been unknown to Darwin, or to have been regarded by him as unconvincing, for even in

---

* 'The Economy of Vegetation,' p. 201.

† Dr. Darwin certainly mentions casually the experiments on *Nicotiana*, by which Kölreuter thought that he had succeeded in converting one plant into another, but he only knew of them from another book.

his last (posthumous) work, 'The Temple of
Nature,' he speaks of the honey-secretion of
plants in the same way as in his earliest writings.   In a special article,* he endeavours to
fathom the secret cause of the general and
abundant secretion of honey by most flowers,
and arrives at the supposition that it is
intended to serve as nutriment and as an
excitant for the sexual organs of the plant,
for which reason it flows only until fertilization has taken place.   He was strengthened
in this curious error by the circumstance
that insects usually go in search of honey
in no other stage of their development
than at the period of their sexual maturity,
that is to say, as perfect insects.   A " philosopher " who seems to have accompanied him
upon this mistaken course, actually supported
his opinion by the absurd conjecture that
the first insects had proceeded from a metamorphosis of the honey-loving stamens and
pistils of the flowers, by their separation
from the parent plant after the fashion of the
male flowers of *Vallisneria*, and " that many

* 'The Economy of Vegetation,' Additional Notes, pp. 107–
112.

" other insects have gradually in long process
" of time been formed from these; some
" acquiring wings, others fins, and others
" claws, *from their ceaseless efforts to procure*
" *their food, or to secure themselves from injury.*
" He (the philosophic friend) contends that
" none of these changes are more incompre-
" hensible than the transformation of tadpoles
" into frogs, and caterpillars into butter-
" flies." *

This error is so instructive and worth
notice, because it shows us the difficulty of explaining a complex natural arrangement, when
one starts from false premises. Could Dr.
Darwin, who afterwards wrote so impressively
upon the mischief of inbreeding, have heard
from any one the magic words " Benefits of
" Cross-fertilization," his error would have
fallen like scales from his eyes; but he firmly
believed that flowers are as far as possible
adapted for self-fertilization, and he stigmatizes
a case of fertilization by the stamens of other
flowers, observed by chance in *Collinsonia,*
with the name of " adultery."† At the same

---

* 'The Economy of Vegetation,' Additional Notes, p. 109.
† 'The Economy of Vegetation,' p. 197, note.

time the exact adaptation of the honey-seek-
ing insects to their business did not escape
him, for in one passage, after describing the
great care which Nature has taken to hide
the honey of the honeysuckle at the bottom of
a long tube (in contrast, incomprehensible to
him, with those flowers in which it lies quite
exposed), he adds that the proboscis of bees
and lepidoptera seems to be especially designed
to reach it in spite of these precautions. "The
" colouring materials of vegetables, like those
" which serve the purpose of tanning, var-
" nishing, and the various medical purposes,
" do not seem," he says in a note on the
madder plant,* " essential to the life of the
" plant; but seem given it as a defence
" against insects or other animals, to whom
" these materials are nauseous or deleterious.
" The colours of insects and many smaller
" animals contribute to conceal them from the
" larger ones which prey upon them. Cater-
" pillars which feed on leaves are generally
" green; earth-worms the colour of the earth
" which they inhabit; butterflies, which fre-
" quent flowers, are coloured like them; small

* 'The Loves of the Plants,' p. 38, note.

" birds which frequent hedges have greenish
" backs like the leaves, and light coloured
" bellies like the sky, and are hence less
" visible to the hawk, who passes under them
" or over them. Those birds which are much
" amongst flowers, as the goldfinch (Fringilla
" carduelis), are furnished with vivid colours.
" The lark, partridge, hare, are the colour of
" dry vegetables or earth on which they rest.
" And frogs vary their colour with the mud of
" the streams which they frequent; and those
" which live on trees are green. Fish, which
" are generally suspended in water, and
" swallows, which are generally suspended in
" air have their backs the colour of the dis-
" tant ground, and their bellies of the sky.
" In the colder climates many of these become
" white during the existence of the snows.
" Hence there is apparent design in the colours
" of animals, whilst those of vegetables seem
" consequent to the other properties of the
" materials which possess them."*

* In the numerous works of the last century which treat of
physico-theology, and especially in those on insecto-theology, in
which the existence of a purpose in all the arrangements of
Nature was discussed in all senses, there are probably numerous
examples of phenomena pertaining to "mimicry." Thus Rösel

In his chief scientific work, the 'Zoonomia,'* to which we now turn, Darwin has also
sought to fathom the causes at work in these
colorations, a matter to which we shall revert
hereafter. The work just mentioned essentially forms a physiology and psychology of
man as a foundation for a pathology, but at
the same time glances are everywhere cast
over the whole animal world. What rank
this work may take in the history of physiology, psychology, and medicine, I cannot
judge, from want of special knowledge in those
departments. Upon the author's contemporaries it produced a very considerable impression, and was immediately translated into
German by a physician of note,† and the translator points out the wonderful agreement of its
views with those of a simultaneously published
work of the celebrated German pathologist

von Rosenhof, in his 'Insekten-Belustigungen" (Nürnberg,
1746), describes the resemblance which the caterpillars of geo-
metric moths, and also certain moths when in repose, present to
dry twigs, and thus conceal themselves, but this group of biological phenomena seems to have been first regarded from a more
general point of view by Dr. Darwin.

* 'Zoonomia, or the Laws of Organic Life.' London, 1794–
1798.

† By Hofrath J. D. Brandis, in 5 vols. Hanover, 1795–1799.

Reil; Hufeland also was strongly influenced by
Darwin. The fundamental idea, it seems to
me, is that in plants and animals a living force
is at work, which, endowed in both with sensibility, is enabled *spontaneously* to adapt them
to the circumstances of the outer world, so
that the assumption of innate ideas, of divinely
implanted impulses and instincts is rendered
unnecessary, and even the process of thought
appears attainable as the legitimate activity
of a mechanical analysis and combination.
All kinds of human knowledge originate from
the senses, the action of which is regarded as
the chief source of knowledge, and is accordingly first of all investigated.

As regards the apparently inborn faculties
which young animals bring with them into the
world, the author explains them by repeated
exertions of the muscles under the guidance
of the sensations and stimuli. Thus it cannot
be wonderful that animals are born into the
world with the faculty of swimming, or of
walking upon four feet, and of swallowing,
for they learnt to swim in the egg or in the
body of the mother, whilst to walk upon two
feet is for quadrupeds an art which does not

belong to nature; the swallowing of fluids is learnt by every fœtus, for they all swallow the amniotic fluid that surrounds them, and it is only the eating of solid matter that requires to be afterwards learned. In the learning of new things the imitative impulse has most to do; and the fact that man, as Aristotle has said, is above all an imitative animal, fits him best for the acquisition of difficult perform-ances,—as, for example, of speech. The author ascribes this desire of imitation even to the smallest constructive parts of the body (as we should say, to the cells), and thereby explains the simultaneous disease of whole complexes of them. The expression of the emotions, also, is acquired by imitation, although their fundamental conditions are organically imposed.

The author very carefully studied this subject, which has been elaborated by his grandson with so much success, and deduces his formulæ especially from the *first* impres-sions of new-born creatures. The trembling of fear may perhaps be referred back to the cold shivering of the new-born infant; and weeping to the first irritation of

the lachrymal glands by cold air, as well
as by pleasant and disagreeable odours.
That anger and rage are universally expressed by animals taking the position of
attack, is immediately intelligible. As regards
smiling and the expression of the agreeable
sensations, the author refers them, as well as
the feeling of the beauty of undulating lines
and of rounded surfaces, to the pleasure of the
first nourishment derived from the soft and
gently rounded maternal breast. "In the
" action of sucking," he says, " the lips of the
" infant are closed around the nipple of its
" mother, till he has filled his stomach, and
" the pleasure occasioned by the stimulus
" of this grateful food succeeds. Then the
" sphincter of the mouth, fatigued by the
" continued action of sucking, is relaxed ; and
" the antagonist muscles of the face gently
" acting, produce the smile of pleasure, as
" cannot but be seen by all who are conver-
" sant with children. Hence this smile during
" our lives is associated with gentle pleasure ;
" it is visible in kittens, and puppies, when
" they are played with and tickled ; but more
" particularly marks the human features.

" For in children this expression of pleasure
" is much encouraged, by their imitation of
" their parents, or friends, who generally
" address them with a smiling countenance :
" and hence some nations are more remark-
" able for the gaiety, and others for the
" gravity of their looks."*

Similarly the wagging of the tails of
animals and the purring of cats are referred
back to certain movements which they acquire
in the time of their existence as sucklings.
" Lambs shake or wriggle their tails, at the
" time when they first suck, to get free of
" the hard excrement which had been long
" lodged in the bowels. Hence this becomes
" afterwards a mark of pleasure in them, and
" in dogs, and other tailed animals. But
" cats gently extend and contract their paws
" when they are pleased, and purr by draw-
" ing in their breath, both which resemble
" their manner of sucking, and thus become
" their language of pleasure, for these animals
" having collar-bones, use their paws like
" hands when they suck, which dogs and
" sheep do not."†   These examples may

* 'Zoonomia,' vol. i. xvi. 8, 4.   † Ib. 8, 3.

serve to show the author's treatment of this difficult theme.

The arts and migratory and social instincts of animals are referred to personal consideration and gradual experience of advantages to be attained. Here also the imitative impulse plays a principal part; and if a horse, for example, wishes to be scratched in a particular part which he cannot reach with his muzzle, he bites his neighbour in the spot in question, and the latter at once understands the hint and does what is required of him. That the arts of animals are acquired is proved by the example already adduced of certain horses stamping down the spiny furze, which the horses of more fertile districts do not understand; and the author also cites many other instances of local deviations and innovations in nest-building and the construction of burrows. Here also we find already mentioned those statements which have been frequently made of late years, with regard to bees which, in certain distant countries (in this case the Island of Barbadoes), store up no honey. The author regards the artificial skill of bees and

ants as very ancient, seeing that it has become
so perfectly developed.

It must not, however, be supposed that the
author regards these instincts as communicated solely by imitation; he accepts without
hesitation the heritability of acquired cor-
poreal peculiarities and mental faculties.
Upon these points there is, in the section
(xxxix.) which treats of generation, and is of
the greatest importance to us, an introductory
observation which contains, as in a nutshell,
the explanation of the biological fundamental
law, and expresses the same ideas which Mr.
Samuel Butler last year made the subject of a
comprehensive book.* " The ingenious Dr.
" Hartley in his work on man, and some
" other philosophers," says Darwin, " have
" been of opinion, that our immortal part
" acquires during this life certain habits of
" action or of sentiment, which become for-
" ever indissoluble, continuing after death in
" a future state of existence; and add, that if
" these habits are of the malevolent kind,
" they must render the possessor miserable
" even in heaven. *I would apply this ingenious*

* 'Life and Habit.' London, 1878.

" *idéa to the generation, or production of the*
" *embryon, or new animal which partakes so*
" *much of the form and propensities of the*
" *parent.*" And he continues as follows:
" Owing to the imperfection of language the
" offspring is termed a *new* animal, but is in
" truth a branch or elongation of the parent ;
" since a part of the embryon-animal is, or
" was, a part of the parent ; and therefore in
" strict language it cannot be said to be
" entirely *new* at the time of its production ;
" and therefore it may retain some of the
" habits of the parent-system."*

It may be observed that the author speaks
here only of one parent ; this is because he
supposed that the embryo consists of the
spermatozoid produced by the father, which
in the mother finds little more than a suitable
nutritive fluid, and a nidus in which it can
develop itself into a perfect animal. The
resemblance of the newly produced creature to
the mother may be explained by the influence
of the nutritive material furnished by her.
Leaving out of consideration this easily ex-
cusable, and, in itself, unimportant error

* ' Zoonomia,' xxxix. 1.

(which I was obliged to mention only because
the author always speaks of a "filament,"
instead of the egg as the germ of the living
creature), the author now, with the greatest
acumen, maintains the theory of epigenesis in
opposition to the theory of evolution (in the
older sense), showing that every creature is a
complete new formation, which, with each grade
of development attained by it, develops other
formative impulses, and thus can incorporate
with its own essence even the latest acquisi-
tions of its parents, by virtue cf the faculty of
recollection possessed by the embryo.   The
old theory of enclosure could not explain such
innovations in the domain of life, and against
it Dr. Darwin therefore turned with lively
sarcasm.   " Many ingenious philosophers,"
he says, " have found so great difficulty in
" conceiving the manner of the reproduction
" of animals, that they have supposed all the
" numerous progeny to have existed in minia-
" ture in the animal originally created; and
" that these infinitely minute forms are only
" evolved or distended, as the embryon in-
" creases in the womb. This idea, besides its
" being unsupported by any analogy we are

" acquainted with, ascribes a greater tenuity
" to organized matter than we can readily
" admit ; as these included embryons are sup-
" posed each of them to consist of the various
" and complicate parts of animal bodies : they
" must possess a much greater degree of
" minuteness, than that which was ascribed
" to the devils that tempted St. Anthony ;
" of whom 20,000 were said to be able
" to dance a saraband on the point of the
" finest needle without incommoding each
" other." *

In the eighth paragraph of the fourth part
of this same section the author gives a short
sketch of the theory of evolution, which, however, must have been more clearly developed
in his mind. I reproduce it here, with some
abridgments, because in it, fifteen years before
the appearance of Lamarck's ' Philosophie
Zoologique,' the principles of evolution were
completely set forth. Darwin says, " When
" we revolve in our minds, first, the great
" changes, which we see naturally produced
" in animals after their nativity, as in the
" production of the butterfly with painted

* ' Zoonomia,' vol. i. § xxxix. iii. 1.

" wings from the crawling caterpillar ; or of
" the respiring frog from the subnatant tad-
" pole ; from the feminine boy to the bearded
" man . . . . . .

 " Secondly, when we think over the great
" changes introduced into various animals by
" artificial or accidental cultivation, as in
" horses, which we have exercised for the
" different purposes of strength or swiftness,
" in carrying burthens or in running races ;
" or in dogs, which have been cultivated for
" strength and courage, as the bull-dog ; or
" for acuteness of his sense of smell, as the
" hound and spaniel ; or for the swiftness of his
" foot as the greyhound ; or for his swimming
" in the water, or for drawing snow-sledges,
" the rough-haired dogs of the north ; . . . . .
" and add to these the great changes of shape
" and colour, which we daily see produced in
" smaller animals from our domestication of
" them, as rabbits, or pidgeons ; or from the
" difference of climates, and even of seasons ;
" thus the sheep of warm climates are covered
" with hair instead of wool ; and the hares and
" partridges of the latitudes which are long
" buried in snow, become white during the

" winter months ; add to these the various
" changes produced in the forms of mankind
" by their early modes of exertion ; or by the
" diseases occasioned by their habits of life ;
" both of which become hereditary, and that
" through many generations. Those who
" labour at the anvil, the oar, or the loom, as
" well as those who carry sedan-chairs, or
" those who have been educated to dance
" upon the rope, are distinguishable by the
" shape of their limbs . . . . . .

" Thirdly, when we enumerate the great
" changes produced in the species of animals
" before their nativity ; these are such as
" resemble the form or colour of their parents,
" which have been altered by the cultivation
" or accidents above related, and are thus
" continued to their posterity. Or they are
" changes produced by the mixture of species,
" as in mules ; or changes produced probably
" by the exuberance of nourishment supplied
" to the fetus, as in monstrous births with
" additional limbs ; many of these enormities
" of shape are propagated, and continued as
" a variety at least, if not as a new species

" of animal. I have seen a breed of cats
" with an additional claw on every foot; of
" poultry also with an additional claw, and
" with wings to their feet; and of others
" without rumps. Mr. Buffon mentions a
" breed of dogs without tails, which are
" common at Rome and at Naples, which he
" supposes to have been produced by a
" custom long established of cutting their
" tails close off. There are many kinds of
" pidgeons, admired for their peculiarities,
" which are monsters thus produced and
" propagated. . . . When we consider all
" these changes of animal form, and innumer-
" able others, which may be collected from the
" books of natural history; we cannot but be
" convinced, that the fetus or embryon is
" formed by apposition of new parts, and not
" by the distention of a primordial nest of
" germs included one within another like the
" cups of a conjurer.

" Fourthly, when we revolve in our minds
" the great similarity of structure which
" obtains in all the warm-blooded animals, as
" well quadrupeds, birds, and amphibious

" animals, as in mankind ; from the mouse
" and bat to the elephant and whale ; one is
" led to conclude, that they have alike been
" produced from a similar living filament.
" In some this filament in its advance to
" maturity has acquired hands and fingers,
" with a fine sense of touch, as in mankind.
" In others it has acquired claws or talons
" . . . in others toes with an intervening
" web, or membrane . . . in others it has
" acquired cloven hoofs . . . and whole hoofs
" in others . . . while in the bird kind this
" original living filament has put forth wings
" instead of arms or legs, and feathers instead
" of hair.   In some it has protruded horns on
" the forehead instead of teeth in the fore part
" of the upper jaw ; in others tushes instead of
" horns ; and in others beaks instead of either.
" And all this exactly is daily seen in the
" transmutations of the tadpole, which acquires
" legs and lungs when he wants them ; and
" loses his tail when it is no longer of service
" to him.

" Fifthly, from their first rudiment, or pri-
" mordium, to the termination of their lives,
" all animals undergo perpetual transforma-

N

" tions, which are in part produced by their
" own exertions in consequence of their
" desires and aversions, of their pleasures and
" pains, or of irritations, or of associations;
" and many of these acquired forms or
" propensities are transmitted to their pos-
" terity.

" As air and water are supplied to animals
" in sufficient profusion, the three great objects
" of desire, which have changed the forms of
" many animals by their exertions to gratify
" them, are those of lust, hunger, and security.
" A great want of one part of the animal
" world has consisted in the desire of the
" exclusive possession of the females; and
" these have acquired weapons to combat
" each other for this purpose, as the very
" thick, shield-like, horny skin on the
" shoulder of the boar is a defence only
" against animals of his own species, who
" strike obliquely upwards, nor are his tushes
" for other purposes, except to defend himself,
" as he is not naturally a carnivorous animal.
" So the horns of the stag are sharp to offend
" his adversary, but are branched for the
" purpose of parrying or receiving the thrusts

" of horns similar to his own, and have,
" therefore, been formed for the purpose of
" combating other stags for the exclusive
" possession of the females; who are observed,
" like the ladies in the time of chivalry, to
" attend the car of the victor.

" The birds which do not carry food to their
" young, and do not therefore marry, are
" armed with spurs for the purpose of fight-
" ing for the exclusive possession of the
" females, as cocks and quails. It is certain
" that these weapons are not provided for
" their defence against other adversaries,
" because the females of these species are
" without this armour. *The final cause of this*
" *contest amongst the males seems to be, that the*
" *strongest and most active animal should pro-*
" *pagate the species, which should thence become*
" *improved.*

" Another great want consists in the means
" of procuring food, which has diversified the
" forms of all species of animals. Thus the
" nose of the swine has become hard for the
" purpose of turning up the soil in search of
" insects and of roots. The trunk of the
" elephant is an elongation of the nose for the

" purpose of pulling down the branches of
" trees for his food, and for taking up water
" without bending his knees.   Beasts of prey
" have acquired strong jaws or talons.   Cattle
" have acquired a rough tongue and a rough
" palate to pull off the blades of grass. . . .
" Some birds have acquired harder beaks to
" crack nuts, as the parrot.   Others have
" acquired beaks adapted to break the harder
" seeds, as sparrows.   Others for the softer
" seeds of flowers, or the buds of trees, as the
" finches.   Other birds have acquired long
" beaks to penetrate the moister soils in search
" of insects or roots, as woodcocks, and others
" broad ones to filtrate the water of lakes, and
" to retain aquatic insects.   *All which seem to*
" *have been gradually produced during many*
" *generations by the perpetual endeavour of the*
" *creatures to supply the want of food, and to*
" *have been delivered to their posterity with*
" *constant improvement of them for the purpose*
" *required.*

" The third great want among animals is
" that of security, which seems much to have
" diversified the forms of their bodies and the
" colour of them; these consist in the means

" of escaping other animals more powerful
" than themselves.* Hence some animals

* The question here only touched upon is discussed in detail
by the author in another part of the ' Zoonomia ' (§ xxxix. 5, 1)
in the following words:—

"The efficient cause of the various colours of the eggs of birds,
and of the hair and feathers of animals, is a subject so curious,
that I shall beg to introduce it in this place. The colours of
many animals seem adapted to their purposes of concealing
themselves, either to avoid danger, or to spring upon their prey.
Thus the snake, and wild cat, and leopard, are so coloured as to
resemble dark leaves and their lighter interstices; birds resemble
the colour of the brown ground, or the green hedges, which they
frequent; and moths and butterflies are coloured like the flowers
which they rob of their honey . . . . . . These colours have,
however, in some instances, another use, as in the black diverging
area from the eyes of the swan; which, as his eyes are placed
less prominent than those of other birds, for the convenience of
putting down his head under water, prevents the rays of light
from being reflected into his eye, and thus dazzling his sight,
both in air and beneath the water; which must have happened,
if that surface had been white like the rest of his feathers.

" There is a still more wonderful thing concerning these
colours adapted to the purpose of concealment; which is, that
the eggs of birds are so coloured as to resemble the colour of the
adjacent objects and their interstices. The eggs of hedge-birds
are greenish, with dark spots; those of crows and magpies,
which are seen from beneath through wicker nests, are white
with dark spots; and those of larks and partridges are russet or
brown, like their nests or situations.

" A thing still more astonishing is, that many animals in
countries covered with snow become white in winter, and are
said to change their colour again in the warmer months. . . .
The final cause of these colours is easily understood, as they
serve some purposes of the animal, but the efficient cause would
seem almost beyond conjecture."

The author endeavoured, however, to clear the way towards

" have acquired wings instead of legs, as the
" smaller birds, for the purpose of escape;
" others great length of fin or of membrane,
" as the flying fish, and the bat.   Others great
" swiftness of foot as the hare.   Others have
" acquired hard or armed shells, as the tortoise
" and the echinus marinus.

" The contrivances for the purposes of
" security extend even to vegetables, as is
" seen in the wonderful and various means of

---

an explanation by saying that the impression of the constant
white light of the snow, or of the yellow of the desert, or of the
green of the woods, might be transferred by reflex action from
the retina to the external papillæ of the skin and its coverings ;
" and thus, like the fable of the camelion, all animals may
possess a tendency to be coloured somewhat like the colours
they most frequently inspect, and finally, that colours may be
thus given to the egg-shell by the imagination of the female
parent." This supposition has lately been proved to be perfectly
correct with respect to certain fishes, amphibia, reptiles, and
mollusca, which always suit themselves to their lighter or darker
surroundings (see Seidlitz, *Die chromatische Funktion als
natürliches Schutzmittel*, in his ' *Beiträge zur Descendenz-
Theorie.*' Leipzig, 1876) ; but it does not suffice for the constant
colorations, notwithstanding the similar hypotheses put forward
by Wallace and others (see Kosmos, iv. p. 120), nor did it by
any means satisfy the elder Darwin, as appears from his further
remarks that the uniformity of the effect would indicate some
other general cause, still to be made out. This cause lies in
natural selection, and the reticence of the elder Darwin in the
face of these circumstances, is the best proof how imperfect any
theory of evolution remains without this principle.

" their concealing or defending their honey
" from insects, and their seeds from birds.
" On the other hand, swiftness of wing has
" been acquired by hawks and swallows to
" pursue their prey ; and a proboscis of
" admirable structure has been acquired by
" the bee, the moth, and the humming bird, for
" the purpose of plundering the nectaries of
" flowers. All which seem to have been formed
" by the original living filament, excited into
" action by the necessities of the creatures,
" which possess them, and on which their
" existence depends.

" From thus meditating on the great
" similarity of the structure of the warm-
" blooded animals, and at the same time of
" the great changes they undergo both before
" and after their nativity ; and by considering
" in how minute a portion of time many of
" the changes of animals above described have
" been produced ; would it be too bold to
" imagine, that in the great length of time,
" since the earth began to exist, perhaps
" millions of ages before the commencement
" of the history of mankind, would it be too
" bold to imagine, that all warm - blooded

" animals have arisen from one living filament
" which THE GREAT FIRST CAUSE endued with
" animality, with the power of acquiring new
" parts, attended with new propensities,
" directed by irritations, sensations, volitions,
" and associations; and thus possessing the
" faculty of continuing to improve by its own
" inherent activity, and of delivering down
" those improvements by generation to its
" posterity, world without end !"

It might be doubted, the author goes on to
say, whether the fishes, which have fins
instead of feet or wings, are of the same
blood as the warm-blooded animals; but
whales, seals, and above all the frog, which
becomes transformed from a fish-like aquatic
animal into an aerial quadruped furnished
with lungs, show that there is no separation
here. On the other hand the insects have
evidently proceeded from a different living
filament, as also the Linnean class of Vermes,
to which sponges, corals, molluscs, &c., were
referred. The same must be supposed with
regard to plants, which the author, like Göthe,
regarded as composite individuals, comparable
to coral stocks.

"Linnæus supposes," continues Darwin,
" in the Introduction to his 'Natural Orders,'
" that very few vegetables were at first created,
" and that their numbers were increased by
" their intermarriages, and adds, *suadent hœc*
" *creatoris leges a simplicibus ad composita.*
" Many other changes seem to have arisen in
" them by their perpetual contest for light
" and air above ground, and for food and
" moisture beneath the soil . . . from climate,
" or other causes. From these one might
" be led to imagine, that each plant at first
" consisted of a single bulb or flower to each
" root, as the gentianella and daisy ; and that
" in the contest for air and light new buds
" grew on the old decaying flower stem,
" shooting down their elongated roots to the
" ground, and that in process of ages tall
" trees were thus formed, and an individual
" bulb became a swarm of vegetables. Other
" plants, which in this contest for light and
" air were too slender to rise by their own
" strength, learned by degrees to adhere to
" their neighbours, either by putting forth
" roots like the ivy, or by tendrils like the
" vine, or by spiral contortions like the honey-

" suckle; or by growing upon them like the
" misleto, and taking nourishment from their
" barks; or by only lodging or adhering on
" them, and deriving nourishment from the
" air, as tillandsia.*

" Shall we then say that the vegetable
" living filament was originally different from
" that of each tribe of animals above-
" described? And that the productive living
" filament of each of those tribes was different
" originally from the other? Or, as the
" earth and ocean were probably peopled with
" vegetable productions long before the exis-
" tence of animals; and many families of
" these animals long before other families of
" them, shall we conjecture that one and the
" same kind of living filaments is and has
" been the cause of all organic life?"

[Here the author refers to the supposition
that America is perhaps the youngest part of

* In his multifarious investigations upon the means of dif-
fusion of the seeds of plants, by wind, flying and projectile contrivances, hooks, fur-animals and birds, he mentions with the
greatest admiration the seeds of *Tillandsia,* which never germi-
nate on the ground. They are provided on their crown with
numerous long filaments, by means of which they fly upon the
winds like spiders, until the threads catch upon the branch of a
tree, and fix the germ there. ('The Loves of the Plants,' p. 60.)

the world, as its inhabitants have not yet ad-
vanced so far in intelligence as those of the
Old World, and its animals (*e.g.* alligators
and tigers) are smaller and weaker. Moreover, the mountains there are higher and less
weathered than ours. That the great lakes of
North America are not yet salt, may be explained by their outflow.]

" This idea of the gradual formation and
" improvement of the animal world," he goes
on to say, " seems not to have been unknown
" to the ancient philosophers. Plato, having
" probably observed the reciprocal generation
" of inferior animals, as snails and worms,
" was of opinion that mankind with all other
" animals were originally hermaphrodites
" during the infancy of the world, and were
" in process of time separated into male and
" female. The breasts and teats of all male
" quadrupeds, to which no use can be now
" assigned, adds perhaps some shadow of
" probability to this opinion. Linnæus ex-
" cepts the horse from the male quadrupeds,
" who have teats; which might have shown
" the earlier origin of his existence; but
" Mr. J. Hunter asserts, that he has discovered

" the vestiges of them . . . . and has at the
" same time enriched natural history with a
" very curious fact concerning the male
" pidgeon ; at the time of hatching the eggs
" both the male and female pidgeon undergo a
" great change in their crops, which thicken
" and become corrugated, and secrete a kind
" of milky fluid, which coagulates, and with
" which alone for a few days they feed their
" young, and afterwards feed them with this
" coagulated fluid mixed with other food.
" How this resembles the breasts of female
" quadrupeds after the production of their
" young! and how extraordinary, that the
" male should at this time give milk as well
" as the female !

    " The late Mr. David Hume, in his posthu-
" mous works, places the powers of genera-
" tion much above those of our boasted
" reason ; and adds, that reason can only
" make a machine, as a clock or a ship, but
" the power of generation makes the maker
" of the machine ; and probably from having
" observed, that the greatest part of the earth
" has been formed out of organic recrements
" . . . . he concludes that the world itself

" might have been generated rather than
" created; that is, it might have been pro-
" duced from very small beginnings, increas-
" ing by the activity of its inherent principles,
" rather than by a sudden evolution of the
' whole by the Almighty fiat.—What a mag-
" nificent idea of the infinite power of THE
" GREAT ARCHITECT! THE CAUSE OF CAUSES!
" PARENT OF PARENTS! ENS ENTIUM!

" For if we may compare infinities, it
" would seem to require a greater infinity of
" power to cause the causes of effects, than to
" cause the effects themselves. This idea is
" analogous to the improving excellence
" observable in every part of the creation;
" such as in the progressive increase of the
" solid or habitable parts of the earth from
" water; and in the progressive increase of
" the wisdom and happiness of its inhabit-
" ants; and is consonant to the idea of our
" present situation being a state of probation,
" which by our exertions we may improve,
" and are consequently responsible for our
" actions."

No one can avoid admitting that in these
considerations, published in 1794, a clear ex-

position is already given of the consequences
of the action of use in its application to the
theory of descent, and therefore of what is
unjustly called Lamarckism. To Lamarck is
to be ascribed the great merit of a further
elaboration of these ideas, but their true
originator and first promulgator appears to
have been the elder Darwin. With the most
perfect certainty we also at the same time
have the principles of a theory of sexual
selection laid down, as far as the consequence
that the strongest male will preferently pro-
pagate, that is to say, within the same limits
in which alone Mantegazza and Wallace are
willing to recognise sexual selection. The
theory of protective coloration is extended to
the eggs of birds, a discovery which has of
late frequently been ascribed to Wallace.
Moreover it deserves to be indicated that
Darwin regards sexual reproduction as a
principal condition of the advancement of
living creatures, as is also the case with
many modern naturalists. It is probable, he
says, " that if vegetables could only have
" been produced by buds and bulbs, and not
" by sexual generation, that there would not

" at this time have existed one thousandth
" part of their present number of species;
" which have probably been originally mule-
" productions; nor could any kind of im-
" provements or change have happened to
" them, except by the difference of soil or
" climate."*

Dr. Darwin believed, moreover, with the
physicians of the last century, that the
imagination of the parents being directed to
certain definite ideals might exert a beneficial
influence upon the young, which would be
impossible in asexual propagation. In a
similar sense the adherents of Geoffroy's
school afterwards thought that the changes of
the world and of the surrounding medium
must have acted more powerfully upon the
plastic embryo than upon the already mature
creature.

A few years after the ' Zoonomia,' Darwin
published his ' Phytologia,'† in which we also
find many coincidences with the investiga-

* ' Zoonomia,' vol. i. xxxix. 6, 2.
† ' Phytologia; or, the philosophy of agriculture and gardening,
with the theory of draining morasses, and with an improved
construction of the drill plough.' London, Johnson, 1800. In
German by Hebenstreit, 2 vols. Leipzig, 1801.

tions of his grandson, especially with regard to artificial selection. Nevertheless we need not here go into it in detail, as his conception of the vegetable world has already been sufficiently explained in its main features in connection with the 'Botanic Garden' and the 'Zoonomia,' whilst some notice will have to be given to it in the consideration of his last work and the general criticism of his system.

'The Temple of Nature,'* dated at the Priory, near Derby, on the 1st of January, 1802, was published in the year following the death of the poet in a quarto volume, adorned, like the 'Botanic Garden,' with fine engravings. It is also a didactic poem, a representation in florid verses of his conception of the universe, fully matured during an interval of ten years. In our rapid analysis we can of course only refer to the novel points of the poem.

In the first canto, which deals with the production of life, &c., we find a decided insistance on the hypothesis of a *Generatio*

---

* 'The Temple of Nature; or, the Origin of Society.' A Poem. London, 1803. In German by Kraus. Brunswick, 1808, 8vo.

æquivoca, the necessity of which he maintains in a note occupying ten quarto pages. In the 'Phytologia' Darwin had set up the hypothesis that the most ancient plants and animals had been destitute of sex, and that the first sexual organs were formed only at a later period. The asexual propagation of many plants and animals, such as the Aphides, which periodically alternates with sexual generation, are reminiscences of this asexual state, and if we then go back still further we arrive necessarily at the hypothesis of spontaneous production :—

> " Hence without parent by spontaneous birth
> Rise the first specks of animated earth."

The examples which he adduces as probable occurrences of spontaneous generation at the present day, such as Priestley's green matter, moulds and other fungi, &c., are certainly not very seductive to an unbeliever, but the acceptance of this hypothesis ought now-a-days to meet with fewer difficulties than that of the rival hypothesis of eternal cosmical life. As a matter of course, as the author remarks, we must only assume spontaneous generation

o

for the simplest creatures of all; all the
higher forms must have been gradually produced from these. This first life originated
in the " shoreless " sea :—

> " Organic life beneath the shoreless waves
> Was born, and nurs'd in ocean's pearly caves ;
> First forms minute, unseen by spheric glass,
> Move on the mud, or pierce the watery mass ;
> These, as successive generations bloom,
> New powers acquire, and larger limbs assume ;
> Whence countless groups of vegetation spring,
> And breathing realms of fin, and feet, and wing."

In the continuation of these verses (lines
295–302) the author recalls to mind that the
higher animals, and even " the image of
God," commence their course of life as microscopic creatures and points :—

> " Imperious man, who rules the bestial crowd,
> Of language, reason, and reflection proud.
> With brow erect who scorns this earthy sod,
> And styles himself the image of his God ;
> Arose from rudiments of form and sense,
> An embryon point, or microscopic ens ! "

Then, when mountains upheaved by the
central fire, or coral reefs, first rose above the
surface of the boundless sea, individual living
organisms landed upon them, and passing

through an amphibious condition, became aerial creatures. " After islands or continents " were raised above the primeval ocean," he says, in a note on p. 29, " great numbers of " the most simple animals would attempt to " seek food at the edges or shores of the new " land, and might thence gradually become " amphibious ; as is now seen in the frog, " who changes from an aquatic animal to an " amphibious one ; and in the gnat, which " changes from a natant to a volant state. " . . . . Those [organisms] situated on dry " land and immersed in dry air, may gradually " acquire new powers to preserve their exist- " ence ; and by innumerable successive repro- " ductions for some thousands, or perhaps " millions of ages, may at length have pro- " duced many of the vegetable and animal " inhabitants which people the earth." As the water-nut (*Trapa natans*) and many other water plants, possess finely divided aquatic leaves, which may be compared with the gills of animals, and also but little divided aerial leaves, comparable to the lungs, so does the frog lose its gills and become instead of a fish-like aquatic animal, an air-breathing quad-

ruped. But even the higher animals in their
embryonic development in the egg or the
body of the mother point towards this origin
from humidity.

> " Still Nature's births enclosed in egg or seed
> From the tall forest to the lowly weed,
> Her beaux and beauties, butterflies and worms,
> Rise from aquatic to aerial forms.
> Thus in the womb the nascent infant laves
> Its natant form in the circumfluent waves ;
> With perforated heart unbreathing swims,
> Awakes and stretches all its recent limbs;
> With gills placental seeks the arterial flood,
> And drinks pure ether from its mother's blood."
>
> (Canto i. l. 385–394.)

In the first canto the poet sings of the
original production of life ; the second has the
" Reproduction of Life " for its subject. In a
note upon this canto a question comes under
discussion for the first time in the works of
the elder Darwin, which his celebrated grandson first settled experimentally, and one of
his great-grandsons (George Darwin) has
made the subject of thorough investigation,
namely the advantage of cross-fertilization
and the mischief of inbreeding.

Dr. Darwin says :—" It may be probably
" useful occasionally to intermix seeds from

" different situations together; as the anther-
" dust is liable to pass from one plant to
" another in its vicinity ; and by these means
" the new seeds or plants may be amended,
" like the marriages of animals into different
" families.

" As the sexual progeny of vegetables are
" thus less liable to hereditary diseases than
" the solitary progenies, so it is reasonable
" to conclude, that the sexual progenies of
" animals may be less liable to hereditary
" diseases, if the marriages are into different
" families, than if into the same family ; this
" has long been supposed to be true, by those
" who breed animals for sale; since if the
" male and female be of different tempera-
" ments, as these are extremes of the animal
" system, they may counteract each other;
" and certainly where both parents are of
" families, which are afflicted with the same
" hereditary disease, it is more likely to
" descend to their posterity. . . . Finally the
" art to improve the sexual progeny of either
" vegetables or animals must consist in
" choosing the most perfect of both sexes,
" that is the most beautiful in respect to the

" body, and the most ingenious in respect to
" the mind ; but where one sex is given,
" whether male or female, to improve a progeny
" from that person may consist in choosing a
" partner of a contrary temperament.   As
" many families become gradually extinct
" by hereditary diseases, as by scrofula,
" consumption, epilepsy, mania, it is often
" hazardous to marry an heiress, as she is
" not unfrequently the last of a diseased
" family."*

His great grandson, George Darwin, has
attempted to demonstrate by statistics these
suppositions, which indeed have been often
expressed, but found that in man no great
injury could be ascertained statistically to
be produced by family marriages, probably
in consequence of the very different conditions under which cousins are frequently
brought up.

We now pass over a hundred verses, and see
what the author has to say in a note on the
Origin of Man.   " It has been supposed by
" some," he says, "that mankind were formerly
" quadrupeds as well as hermaphrodites ; and

* ' Temple of Nature,' Additional Notes, pp. 44, 45.

" that some parts of the body are not yet so
" convenient to an erect attitude as to a hori-
" zontal one; as the fundus of the bladder
" in an erect posture is not exactly over the
" insertion of the urethra; whence it is sel-
" dom completely evacuated, and thus renders
" mankind more subject to the stone, than if
" he had preserved his horizontality; these
" philosophers, with Buffon and Helvetius,
" seem to imagine that mankind arose from
" one family of monkeys on the banks of the
" Mediterranean; who accidentally had learned
" to use the adductor pollicis, or that strong
" muscle which constitutes the ball of the
" thumb, and draws the point of it to meet
" the points of the fingers; which common
" monkeys do not; and that this muscle gradu-
" ally increased in size, strength, and activity,
" in successive generations; and by this im-
" proved use of the sense of touch, that
" monkeys acquired clear ideas, and gradually
" became men."*

The great part performed by the hand and
its improved sense of touch is specially described in the third canto, which is devoted

* 'Temple of Nature,' note p. 54.

to the development and progress of the human
mind.   Animals excel man in being endowed
with many kinds of weapons and in having
the senses more highly developed, but the
influence of the hand in forming the mind
more than compensates for all :—

> " Proud man alone in wailing weakness born,
>   No horns protect him, and no plumes adorn;
>   No finer powers of nostril, ear, or eye,
>   Teach the young Reasoner to pursue or fly.—
>   Nerved with fine touch above the bestial throngs,
>   The hand, first gift of Heaven! to man belongs;
>   Untipt with claws the circling fingers close,
>   With rival points the bending thumbs oppose,
>   Trace the nice lines of Form with sense refined,
>   And clear ideas charm the thinking mind.
>   Whence the fine organs of the touch impart
>   Ideal figure, source of every art;
>   Time, motion, number, sunshine or the storm,
>   But mark varieties in Nature's *form*."
>   (Canto iii. l. 117–130.)

In young dogs, adds the author, the
lips are the principal organs which enable
them to acquire an idea of the forms of
things; and in young children also the lips
play a great part in the same way. He then
describes very fully the functions of the impulse of imitation in man, attributing to it

the first origin of all moral actions, languages and arts.

The " Muse of Mimicry," as Darwin, in what follows, repeatedly calls the love of imitation in man, gave rise especially in his opinion to the first language, and the first writing, which was a picture-writing.

On the problem of the origin of language, the learned Lord Monboddo's anonymously published work ('Of the Origin and Progress of Language'), in three volumes, had at that time been in existence for a quarter of a century. In this book he shows, by the study of animals and deaf-mutes, in opposition to recent observers, that without speech it is possible to think and to form ideas, for speechless animals manifestly have ideas (vol. i. p. 217 *et seq.*). Signs and musically varied cries formed the commencement (i. p. 476). Articulation was acquired by imitation of natural sounds, e.g. the voices of birds (v. p. 490). Even such new conceptions as that of A. Maurer, that primitive speech was formed not by monosyllabic but polysyllabic words, are already to be found here (i. pp. 507 *et seq.*). It is decidedly much to be wished

that some philologist would analyse this forgotten book from the modern standpoint. We
are not sufficiently skilled to examine how
much novelty there may be in Dr. Darwin's
studies in this department, and must content
ourselves with calling attention to the detailed
considerations which he gives in his notes, in
connection with which we may here reproduce
one of the most characteristic passages in the
poem :—

> " When strong desires or soft sensations move
> The astonished Intellect to rage or love ;
> Associate tribes of fibrous motions rise,
> Flush the red cheek, or light the laughing eyes.
> Whence ever-active Imitation finds
> The ideal trains, that pass in kindred minds ;
> Her mimic arts associate thoughts excite
> And the first LANGUAGE enters at the sight."
>
> (Canto iii. 1. 335–342.)

After showing how true language has
originated from the language of the emotions
and gestures, from the first exclamations,

> (" Association's mystic power combines
> Internal passions with external signs.")

he traces the accentuation and articulation
of sounds, the formation of fundamental
words and abstract ideas, the growth of

intellect intimately connected with these pro-
cesses, and the origin of the social virtues or
general morality founded upon social inter-
course. The fundamental principle of the
latter is best expressed in the words of Christ,
" Love thy neighbour as thyself."

. The fourth canto, entitled "Of Good and
Evil," represents the spiritual as a stage of
development of the material world, the sum of
the happiness. and evil therein. About the
first hundred verses are devoted to a description of the pitiless struggle for existence
which rages in the air, on the earth, and in
the water, making the earth, with its incessantly warring inhabitants, like a vast
slaughter-house :—

> " Air, earth, and ocean, to astonish'd day
> One scene of blood, one mighty tomb display!
> From Hunger's arm the shafts of Death are hurl'd,
> And one great Slaughter-house the warring world!"
>
> (Canto iv. l. 63–66.)

This description is no mere passing notion,
for in his first didactic poem, ' The Botanic
Garden,' written at least twenty years
before, this same idea occurs (p. 28). Dr.
Balguy had indicated the benefits with which
the great Author of all things had favoured

the world.   The young animal takes the
mother's breast with pleasure, and the mother
has pleasure in offering it.   The seeds of
plants, rich in nutritive material, serve animals
for food without themselves feeling pain.
Against this much too rose-coloured conception
of the world, our author protested at the time.
The lion devours the lamb, and the latter
the living plants, whilst man *eats* both; there
is nothing like peace in nature.   In his last
work this conception appears to have been
much deepened; not only do animals destroy
each other and plants, but even the plants
struggle among themselves for soil, moisture,
air, and light:—

" Yes! smiling Flora drives her armed car
    Through the thick ranks of vegetable war;
    Herb, shrub, and tree with strong emotions rise
    For light and air, and battle in the skies;
    Whose roots diverging with opposing toil
    Contend below for moisture and for soil;
    Round the tall Elm the fluttering Ivies bend,
    And strangle, as they clasp, their struggling friend;
    Envenom'd dews from Mancinella flow,
    And scald with caustic touch the tribes below;
    Dense shadowy leaves on stems aspiring borne
    With blight and mildew thin the realms of corn;
    And insect hordes with restless tooth devour
    The unfolded bud, and pierce the ravell'd flower."

                                    (Canto iv. l. 41–54.)

Fortunately living creatures often struggle
with each other for the advantage of a third
party, as when the voracious larvæ of insects
which, after their metamorphosis, live only
on honey, destroy the innumerable hosts of
aphides, which otherwise, from their enormous fertility would exterminate all vegetation. An excess of the caterpillars of butterflies is destroyed by hymenopterous insects ;
moreover plants are able to protect themselves from complete destruction. Nevertheless
this never-resting struggle of all against all,
would soon create desolation, if Nature was
not so exceedingly fruitful that without such
a struggle nearly every creature would very
soon overrun the whole world :—

> " All these, increasing by successive birth,
> Would each o'erpeople ocean, air, and earth."

Here is the great question put, What is the
meaning for Nature, of this incessant struggle
in Nature ? For a moment we may perhaps
expect to get the solution of this mystery of
Nature from the poet who had come so near
to it, but it is only a presentiment of the truth,
not the truth itself. Thus he says that the

incessant struggle serves to increase the sum
of the happiness of the survivors :—

> " Thus the tall mountains, that enclose the lands,
>   Huge isles of rock, and continents of sands,
>   Whose dim extent eludes the inquiring sight,
>   ARE MIGHTY MONUMENTS OF PAST DELIGHT ;
>   Shout round the globe, how Reproduction strives
>   With vanquish'd Death,—and HAPPINESS SURVIVES ;
>   How life increasing peoples every clime,
>   And young renascent Nature conquers Time ;
>   —— And high in golden characters record
>   The immense munificence of NATURE'S LORD."
>
> (Canto iv. l. 447–456.)

By the increased happiness which arises from
the death of those which fall in the struggle,
the author, however, chiefly understands that
fresh life blooms from dull age, and that, as
both the number and the size of living animals
increase with the decrease of the water, the
sum of enjoyment of life must also increase,
until the earth is once more reduced to its
elements, in order, through chaos, to commence a new cycle.* The principle of the
reconversion of the world into chaos, also
supported by modern physics, is laid down by
the author in his ' Botanic Garden ' with

---

* ' Temple of Nature,' p. 166 note.

such force that I cannot refrain from giving
this passage as a final example of his poetical
power :—

> " —— Roll on, ye Stars! exult in youthful prime,
> Mark with bright curves the printless steps of time;
> Near and more near your beamy cars approach,
> And lessening orbs on lessening orbs encroach;
> Flowers of the sky! ye too to age must yield,
> Frail as your silken sisters of the field!
> Star after star from Heaven's high arch shall rush,
> Suns sink on Suns, and systems systems crush,
> Headlong, extinct, to one dark centre fall,
> And Death, and Night, and Chaos mingle all!
> —— Till o'er the wreck, emerging from the storm,
> Immortal NATURE lifts her changeful form,
> Mounts from her funeral pyre on wings of flame,
> And soars and shines, another and the same."

In his 'Phytologia' (xix. 7) the author
has treated still more in detail the question
of the struggle for existence, and the sum of
happiness originating therefrom, and he indi-
cates in the note last cited that the faculty of
higher enjoyment increases with the height of
organization of the creatures. He had not
indeed solved the question, but his remarks
upon it have directed the eyes of many of
his readers to the struggle for existence, and
in this we may perhaps find the explanation
of the remarkable fact that so many English

naturalists (Wells, Matthew, Charles Darwin, Wallace, &c.) have one after the other set up the principle of natural selection. This shows the power of the poet to excite the fancy even of others; and a happy fate has arranged that the true heir has obtained the greatest benefit from the bequest.

The ' Temple of Nature ' contributed greatly to enhance Darwin's poetical fame, for the representation is more rounded, and less over-grown with allegorical comparisons, than in his first didactic poems. But how little the philosophy expressed in it satisfied the readers of that time may be seen from a criticism which was given of the poem in the ' Edin-burgh Review ' (vol. ii. 1803, pp. 491–506). In it occurs (p. 501) the following remark, which is interesting in two ways :—" If his " fame be destined in anything to outlive " the fluctuating fashion of the day, it is on " his merit as a poet that it is likely to rest; " and his reveries in science have probably " no other chance of being saved from " oblivion, but by having been ' married to " immortal verse.' "

This full recognition of the author's poetical

merits contrasts curiously enough with the sharp judgment of a later critic,* who, I am afraid, has criticized himself in it. "Nothing "in them," he says of the verses, "is done in "passion and power; but all by filing, and "scraping, and rubbing, and other pains- "taking. Every line is as elaborately "polished and sharpened as a lancet; and "the most effective paragraphs have the air "of a lot of those bright little instruments "arranged in rows, with their blades out, for "sale. You feel as if so thick an array of "points and edges demanded careful handling, "and that your fingers are scarcely safe in "coming near them." We see at once that the critic cannot forgive the poet for having been a doctor; regards thought as a me- chanical process, and poetry as mechanical work, a higher kind of "pin-making." After the critic has thus shot his arrows, however, he is obliged to admit that in spite of all a true poetical fire lives in these didactic poems and frequently breaks forth. "No writer,"

* George L. Craik. 'A Compendious History of English Literature, and of the English Language, from the Norman Conquest.' 2nd ed. vol. ii. pp. 382, 383. 8vo. London, 1864.

says he, " has surpassed him in the luminous
" representation of visible objects in verse ;
" his descriptions have the distinctness of
" drawings by the pencil, with the advantage
" of conveying, by their harmonious words,
" many things that no pencil can paint."

We will be more just, and say, that since
the time of Lucretius, hardly any attempt to
combine the opposing spheres of science and
poetry in a didactic poem, and to put forth
therein entire systems, has been so successful
as in Darwin's works ; but such poems are
rather dry in themselves, and will always find
fewer admirers than poetical efforts of other
kinds. Nevertheless even if the body of these
poems should prove to be mortal, an immortal
spirit lives in them, and it is this (to turn the
words of the Edinburgh Reviewer the other
way round) that will keep them above water
for all time.

Now, at the conclusion of our analysis, it
may be as well to take a general view of the
system established by Dr. Erasmus Darwin,
in order to arrive at a clear perception of the
advance for which the conception of the
universe is indebted to him, and of the points

in which he erred. And here we must in
the first place admit *that he was the first who
proposed and consistently carried out, a well-
rounded theory with regard to the development
of the living world*, a merit which shines forth
most brilliantly when we compare with it the
vacillating and confused attempts of Buffon,
Linnæus and Göthe. It is the idea of a power
working from within the organisms to improve their natural position; and thus, out of
the impulses of individual needs, to work towards the perfection of Nature as a whole.

In contrast to the old theory that all adaptation to purpose in the arrangements of the
world was fore-calculated and fore-ordained,
and that all organisms were merely wheels in
a gigantic machine made once for all, and in-
capable of improvement, this new view is so
grand that it deserved a higher appreciation
than it has ever met with. The Cartesio-
Paleyan comparison of Nature with a great
piece of clockwork (a fundamentally mistaken
comparison, because every complete mechanical work has only been attained by many
gradual improvements in the course of generations), is finally got rid of by it. As regards

the animal world, to which we must ascribe
will and active efforts, the idea is so suitable,
that Lamarck, who was evidently a disciple
of Darwin, has worked it out in all directions,
and thus originated a system which is not
only still appreciated, but is even now constantly being further elaborated, inasmuch as
many naturalists of the present day, as has
already been stated, ascribe to birds, for example, the faculty of enhancing the beauty of
their plumage by wishes and efforts, and so
forth. This is true Darwinism of the last
century—Darwinism of the old school.

This Darwinism has been criticized by
no one so well as by its author himself,
when he applied it with strict logic to the
development of plants. To be able to do this
he was obliged to attribute mental functions
to plants, and to endow them with the faculty
of striving for a purpose. Even in the
'Botanic Garden' he therefore declared the
necessity of admitting that plants possess the
sense of heat and cold, of moisture and dry-
ness, of light and darkness, a sense of touch,
and amatory desires, besides the power of the
roots to select suitable nourishment. For

these reasons he also specially occupied himself with the study of the so-called sensitive plants, and of insect-capturing plants, the most remarkable of which (*Mimosa, Hedysarum gyrans, Dionœa muscipula, Apocynum androsœmifolium*) he had figured on fine quarto plates to illustrate the 'Botanic Garden.'

In the 'Zoonomia' he repeated these views; and in the first part of the 'Phytologia,' which treats of the physiology of plants, he is much occupied with the search for vegetable organs representing the organs of sense, nerves, and ganglia, of animals. Nay, he even thought that an organ analogous to the central nervous apparatus of animals, a vegetable brain, could not be wanting; and as he rightly compared the composite vegetable body to a coral-stock, he was obliged to ascribe such an organ to each individual bud. For as he ascribed to them (and according to his theory was compelled to ascribe to them), besides the power of nourishing and propagating themselves, also that of endeavouring to improve their position in life in accordance with external conditions, he logically concluded that he must for this purpose postulate an organ of self-help, a sensorium.

In order to penetrate more clearly into the course of his ideas upon this point, I may be allowed to quote, in part, a passage from the 'Phytologia' (Sect. xiv. 3, 2), and the rather because it at the same time fills a gap purposely left in the exposition of his philosophical system.

" There appears," he says, " to be a power " impressed on organized bodies by the great " author of all things by which they not only " increase in size and strength from their " embryon state to their maturity, and oc- " casionally cure their accidental diseases, and " repair their accidental injuries, but also a " power of producing armour to prevent those " more violent injuries, which would other- " wise destroy them. Of this last kind are " the poisonous juices of some plants, as of " *atropa belladonna*, deadly nightshade, *hy-* " *oscyamus*, hen-bane, *cynoglossum*, hound's " tongue . . . . . . Some vegetables have " acquired an armour, which lessens, though " it does not totally prevent, the injuries of " this animal [the aphis]. This is most con-

" spicuous on the stems and floral leaves of
" moss-roses, and on the young shoots and
" leaf-stalks of nut trees. Both these are
" covered with thickset bristles, which termi-
" nate in globular heads, and not only prevent
" the aphis from surrounding them in such
" great numbers, and from piercing their
" vessels so easily, but also secrete from the
" gland, with which I suspect them to be
" terminated, a juice, which is inconvenient or
" deleterious to the insect which touches it.*
" . . . The essential oils are all deleterious to
" certain insects, and hence their use in the
" vegetable economy, being produced in
" flowers or leaves to protect them from the
" depredations of their voracious enemies."

I do not think I am deceiving myself in
saying that this merely logical extension of
his theory to the vegetable kingdom has
robbed it of the efficacy which it might have
attained if limited to the animal kingdom.
The small amount of interest excited by the
attempts, both of the elder Darwin and of

* Corresponding observations upon glandular hairs which, by
their sticky exudations, protect young shoots of plants from the
attacks of insects, have lately been made by Mr. Francis Darwin
and by Dr. Fritz Müller. See ' Kosmos,' Bd. i. p. 354.

Lamarck, at the solution of the world-enigma, shows us that they were not adapted to satisfy men's minds. They explain the adaptation to purpose of organisms by an obscure impulse or sense of what is purpose-like; and yet even with regard to man we are in the habit of saying, that one can never know what so and so is good for. *The purpose-like is that which approves itself*, and not always that which is struggled for by obscure impulses and desires. Just in the same way *the beautiful is what pleases*.

Erasmus Darwin's system was in itself a most significant first step in the path of knowledge which his grandson has opened up for us, but to wish to revive it at the present day, as has actually been seriously attempted, shows a weakness of thought and a mental anachronism which no one can envy.

Printed in the United States
By Bookmasters